This book is to be returned on or before
the last date stamped below.

IIN 1991

# Solidification Processing of Eutectic Alloys

# Solidification Processing of Eutectic Alloys

Proceedings of a symposium sponsored by the TMS Solidification Committee and the Powder Metallurgy Committee held at the TMS Fall Meeting, October 12-15, 1988, Cincinnati, Ohio.

Edited by

**D.M. Stefanescu**
*The University of Alabama*
Tuscaloosa, AL

**G.J. Abbaschian**
*University of Florida*
Gainesville, FL

**R.J. Bayuzick**
*Vanderbilt University*
Nashville, TN

D
669.94
SOL

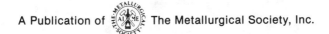
A Publication of The Metallurgical Society, Inc.

**A Publication of The Metallurgical Society, Inc.**
420 Commonwealth Drive
Warrendale, Pennsylvania 15086
(412) 776-9000

Printed in the United States of America.
Library of Congress Catalogue Number 87-43113
ISBN NUMBER 0-87339-033-4

# PREFACE

The eutectic transformation plays a major role during solidification processing of a variety of alloys.  In fact, about half of the existing commercial alloys undergo some type of eutectic transformation during their processing, and thus the properties of these alloys strongly depend on the amount and morphology of the eutectic phases, which, in turn, are affected by processing variables.  Among these variables one must include cooling rate, nucleation hierarchy and the faceted or non-faceted nature of the constituent phases.  This book addresses the effect of the above mentioned variables on the eutectic morphology and resulting properties.  The volume is a record of the proceedings of the Conference on Solidification Processing of Eutectic Alloys which was held in Cincinnati, Ohio at the 1987 Fall Meeting of The Metallurgical Society.  The Conference attracted twenty invited and contributed papers out of which fifteen are published in this book.

A critical evaluation of the current status of the modeling of eutectic growth, to include the influence of convection on eutectic morphology, as well as computer modeling of microstructure evolution of eutectics was addressed in a number of papers.  Directional solidification experiments of a variety of alloys such as aluminum foundry alloy A 356, Si- and GaAs-based eutectics for electronic applications, and Cu-Al in-situ composites were also presented.

The most widely used eutectic alloy at the present time is still cast iron, and three papers dealing with this material discussed modeling of solidification, microstructure and residual stress.

Particular attention was given to undercooled and rapidly solidified alloys, including classical, simple alloys such as Pb-Sn as well as more complicated alloys such as Nb-Si, Alumina-Zirconia and $Cr_{90}Ta_{10}$.  The techniques used for rapid solidification included melt spinning, argon plasma sputtering, electrohydrodynamic atomization, direct-energy electron-beam processing and electromagnetic levitation.

The success of the Conference was made possible not only by the active participation of the contributors, but also by the extra effort involved in preparing the invited papers as well as chairing the session.  We are grateful to B. MacDonald, D.A. Granger and R. Trivedi for their active involvement in the conference as chairmen and/or invited speakers.

Finally, we wish to offer sincere thanks to Mrs. Donna Snow for secretarial services throughout the lengthy task of producing this volume.

D.M. Stefanescu
The University of Alabama
Tuscaloosa, AL

G.J. Abbaschian
University of Florida
Gainesville, FL

R.J. Bayuzick
Vanderbilt University
Nashville, TN

# TABLE OF CONTENTS

vii

# DIRECTIONAL
# SOLIDIFICATION

MICROSTRUCTURE SELECTION IN EUTECTIC ALLOY SYSTEMS

R. Trivedi[*] and W. Kurz[†]

[*]Ames Laboratory, USDOE and the Department of
Materials Science and Engineering
Iowa State University
Ames, Iowa 50011

†Department of Materials Engineering,
Swiss Federal Institute of Technology,
Lausanne, Switzerland

Abstract

The current status of the modeling of eutectic growth is critically examined. Relevant experimental studies are described and it is shown that the operating point of eutectic spacing deviates from the minimum undercooling value. This deviation is found to be small for regular eutectics, whereas it is significantly larger for irregular eutectics. At low velocities, the average eutectic spacing is inversely proportional to the square root of velocity. However, this relationship is altered at high growth rates. It is shown that eutectic microstructures do not form above some critical velocity and the reasons for this behavior are discussed. The conditions under which a eutectic structure will be preferred over dendritic or cellular structure are developed and the important parameters which give rise to regular and skewed coupled zones are discussed. The various microstructural selections as a function of undercooling are represented on a phase diagram to create microstructural maps which visually exhibit various possible stable and metastable microstructures in a given alloy system under normal and rapid solidification conditions.

Solidification Processing of Eutectic Alloys
D.M. Stefanescu, G.J. Abbaschian and R.J. Bayuzick
The Metallurgical Society, 1988

# I. Introduction

Most solidification microstructures can be broadly classified into two important growth morphologies: dendritic and eutectic microstructures. If we compare the characteristic scales of these two microstructures, the eutectic spacing and the dendrite tip radius have nearly the same magnitude for given conditions. However, under normal solidification conditions, the primary and the final secondary dendrite spacings are about an order of magnitude larger than the dendrite tip radius. Consequently eutectic microstructures are often finer so that they exhibit superior mechanical properties. Furthermore, eutectic structures can be directionally solidified to form in situ composites. When the second phase is strong, as in the case of TaC-fibers in the Ni-TaC eutectic system, a marked increase in the creep strength of the alloy is found.

Eutectic alloys have lower melting points compared with those of the pure components. Furthermore, in spite of the large alloy content, the small freezing range of these alloys gives them excellent flow properties. These properties of eutectic alloys make them extremely valuable in casting, welding and soldering processes. In fact, these properties have led to the word eutectic, which is of Greek origin and means 'more fusible'.

It is important to first distinguish between a eutectic equilibrium and a eutectic microstructure. In a eutectic system, a unique temperature ($T_E$) and composition ($C_E$) exist at which the liquid is in equilibrium with the different solid phases, as shown in Fig. 1. Eutectic microstructures, on the other hand, are described by the cooperative growth of two phases from the liquid in case of binary alloys. These two phases form either a lamellar or a rod structure, as shown in Fig. 2. Eutectic alloys

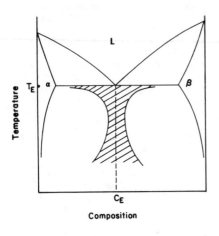
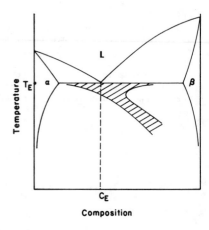

(a)                                    (b)

Fig. 1. Schematic phase diagrams which show the eutectic temperature, $T_E$, and the eutectic composition, $C_E$. The shaded regions are the temperature-composition zones in which a coupled eutectic growth occurs. (a) Regular coupled zone. (b) Skewed coupled zone.

4

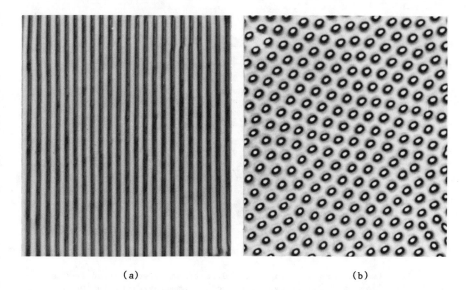

(a)        (b)

Fig. 2. Transverse cross-sections of directionally solidified alloys which
illustrate (a) a lamellar structure in the Pb-17.4 wt% Cd alloy and
(b) a rod structure in the Sn-18.0 wt% Pb system.


do not always form eutectic microstructures, especially at larger undercoolings. Also a eutectic microstructure can be formed for alloys which are far
away from the eutectic composition. Fig. 1 shows the composition - temperature zone in which eutectic microstructures can be obtained, and this region
is generally referred to as the coupled zone.

From a technological viewpoint, there are two important parameters of
eutectic microstructures which can be controlled experimentally to influence
the properties of the material. These are: (1) the eutectic spacing and
(2) the volume fractions of the two phases. Eutectic spacing is primarily
controlled by the growth rate*, whereas the volume fractions are controlled
by the composition of the alloy . In order to discuss eutectic microstructures, we shall examine the following microstructural characteristics:

(1) The relationship between the eutectic spacing and the growth rate. We
shall first examine this relationship for small growth rate conditions,
and then show how this relationship changes under rapid solidification
conditions. Specifically, we shall discuss the highest velocity, and
the smallest spacing which give stable eutectic structures.

(2) The effect of composition on the stability of eutectic structures, i.e.
the range of coupled growth. This coupled zone can be understood
through the relative growth kinetics of the eutectic and the dendritic
structures. Based on these kinetic considerations, we shall discuss the
microstructure selection criterion in a given alloy system.

---

*In this paper we shall consider directional solidification only. For an
undercooled melt, the control parameter is undercooling which determines
the growth rate.

5

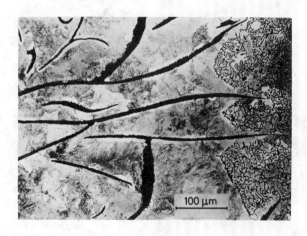

Fig. 3. Longitudinal cross-section of a directionally solidified Fe-C eutectic alloy, showing irregular eutectic lamellae distribution and the quenched nonplanar solid-liquid interface at right.

The above characteristics will be considered for regular as well as for irregular eutectics. In a regular eutectic, the interface is nearly planar so that both phases have the same average undercooling at the interface. On the other hand, in an irregular eutectic, shown in Fig. 3, the two phases have different average undercoolings and one of the phases strongly controls the overall growth.

The basic problem of eutectic growth at low velocities will be considered in section II, where the diffusion problem and the stability of eutectic microstructures will be discussed to gain some insight into the operating point of the eutectic microstructure. These theoretical ideas will then be examined in light of critical experimental studies in metallic and organic systems. In section III, the important developments in eutectic microstructure at high growth rates will be discussed. Specifically, the relationship between the growth rate and the eutectic spacing will be established and the reasons why eutectic structures cannot be obtained above a certain critical velocity will be presented. Finally, in section IV, major ideas behind the nature of the coupled zone will be outlined and the effect of growth rate and composition on stable and metastable phase selection will be discussed.

## II.  Eutectic Growth at Low Velocities

The eutectic microstructure requires simultaneous growth of two phases and this is possible if the lateral transfer of solute through liquid between the two phases is easier than the diffusion in liquid in the direction of growth. Consequently, diffusion plays a critical role in determining microstructural characteristics of eutectic structures. However, the diffusional considerations alone would cause the spacing to be very fine. In contrast, as the spacing becomes finer, more surface area between the two phases is created for a unit volume of transformation. Thus, the key aspects of theoretical model of eutectic growth involve a balance between

6

the diffusion and the surface energy considerations. These two factors, however, give rise to multiple solutions so that an additional criterion is required which determines the selection of a specific spacing under given experimental conditions. In this section, we shall first outline the diffusional problem and then discuss whether the stability analysis of eutectic structures can uniquely determine the spacing. Finally, we shall examine critical experimental studies which give valuable information about the operating point of eutectic spacings.

## II-1. Diffusion-Coupled Growth

A detailed theoretical model of regular eutectic growth was developed by Jackson and Hunt (JH) [1]. For two solid phases, $\alpha$ and $\beta$, they calculated the average undercooling at the $\alpha$/liquid and the $\beta$/liquid interfaces by first partitioning the total interface undercooling, $\Delta T_i$, for each phase into the following components,

$$\Delta T_i = \Delta T_S + \Delta T_c, \quad i = \alpha \text{ or } \beta, \quad (1)$$

where $\Delta T_S$ and $\Delta T_C$ are the solutal and the capillary undercoolings, respectively.

The solutal undercooling was obtained by solving the steady-state diffusion equation which, in a coordinate system attached to the interface, is given by

$$\nabla^2 C + (V/D)(\partial C/\partial Z) = 0, \quad (2)$$

where the Z-coordinate is along the growth direction and V and D are the growth rate and the diffusion coefficient in liquid, respectively. The capillary undercooling was obtained by using the Gibbs Thompson equation, which is

$$\Delta T_C = \Gamma \kappa, \quad (3)$$

where $\Gamma = \gamma/\Delta S$, in which $\gamma$ and $\Delta S$ are the interfacial free energy and the entropy of fusion, respectively. $\kappa$ is the average curvature of the interface for a given phase.

Jackson and Hunt solved equation (2) for low undercooling values only. In order to assess the conditions under which their model is valid, we shall briefly examine some of the assumptions made in their treatment.

(1) The diffusion distance ahead of the interface, D/V, was assumed to be large compared to the eutectic spacing, $\lambda$. This assumption allowed them to ignore the last term on the left hand side of equation (2). Thus, they obtained the diffusion field for a stationary interface by solving the Laplace equation.

(2) Laplace equation for the diffusion field was solved by assuming that the eutectic front is planar, see Fig. 4 (lower part). The average curvature of the lamella or the rod was considered only for determining the average capillary undercooling at the interface.

7

(3) The undercooling was considered to be small so that the compositions of α and β phases at the interface were independent of undercooling. The values of compositions, $C_\alpha$ and $C_\beta$, were taken to correspond to the eutectic temperature.

(4) The average interface undercoolings for the two phases were assumed to be equal, i. e. $\Delta T_\alpha = \Delta T_\beta = \Delta T$.

Under these assumptions, JH obtained the relationship between the undercooling, growth rate and eutectic spacing, as

$$\Delta T = K_1 \lambda V + K_2/\lambda, \qquad (4)$$

where $K_1$ and $K_2$ are system parameters, which have the following values,

$$K_1 = mPC_o/f_\alpha f_\beta D \qquad (5)$$

and

$$K_2 = 2m\delta \sum_i (\Gamma_i \sin\theta_i/m_i f_i); \quad i = \alpha, \beta, \qquad (6)$$

where $m = m_\alpha m_\beta/(m_\alpha + m_\beta)$ in which $m_\alpha$ and $m_\beta$ are the magnitudes of the α and β liquidus slopes at the eutectic temperature, $C_o$ is the difference between the composition in the β and the α phase, $f_\alpha$ and $f_\beta$ are the volume fractions of α and β phases, respectively, and the angles $\theta_i$ are shown in Fig. 4. The parameter δ is unity for the lamellar growth and it is equal to $2\sqrt{f_\alpha}$ for the rod eutectic growth. The parameter P is defined as

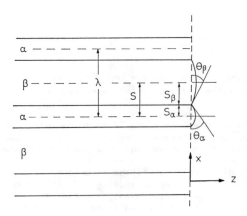

Fig. 4. Schematic diagram of eutectic structure which defines the coordinate system and the contact angles at the triple point junctions.

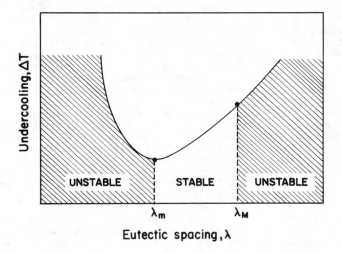

Fig. 5. Relationship between the average interface undercooling and the eutectic spacing at a fixed velocity, and the regions of stable and unstable spacings, as predicted by the JH model [1].

$$P = \begin{cases} \sum\limits_{n=1}^{\infty} \dfrac{1}{(n\pi)^3} \, \mathrm{Sin}^2(n\pi f_\alpha) & \text{for lamellar eutectic} \quad (7a) \\[4mm] 2f_\alpha \sum\limits_{n=1}^{\infty} \dfrac{1}{(\gamma_n)^3} \, \dfrac{J_1^2 \, (\gamma_n f_\alpha)}{J_o^2 \, (\gamma_n)} & \text{for rod eutectic} \quad (7b) \end{cases}$$

$J_1$ is the Bessel function of first order and $\gamma_n$ are the roots of $J_1(x) = o$ which are approximately equal to $n\pi$. The numerical values of P for lamellar and rod eutectics have been tabulated by JH*. These functions can be written in a simplified form as

$$P = \begin{cases} 0.3383(f_\alpha f_\beta)^{1.661}, & \text{for lamellar eutectic} \quad (8a) \\[4mm] 0.167(f_\alpha f_\beta)^{1.25}, & 0 \le f_\alpha \le 0.3 \quad \text{for rod eutectic} \quad (8b) \end{cases}$$

The general solution of the diffusion problem, given by equation (4), predicts that the eutectic spacing at a given growth rate depends on undercooling, as illustrated schematically in Fig. 5. Jackson and Hunt discussed that all these solutions are not stable with respect to small changes in the interface shape, and they qualitatively discussed the range of stable spacings, as shown in Fig. 5.

---

*For rod eutectic, JH have given values of constant M, which is related to our definition as follows: $P(\text{rod}) = 2f_\alpha M$.

Before we proceed to discuss the operating point of the eutectic spacing, we shall briefly examine the range of validity of the transport solution of the JH model. (a) The use of Laplace equation instead of the complete equation (2) is valid only when $\lambda \ll D/V$. A detailed analytical treatment by Trivedi et al. [2] which considered the complete diffusion equation showed that the JH result is valid for metals up to growth rates of about 10 mm/s. (b) Series et al. [3] have used an electrical analogue to examine the effect of change in solid compositions with undercooling. They observed that $V\lambda^2$ deviates from a constant value as the undercooling is increased. This deviation occurs due to the change in $C_0$ which causes $K_1$ to change with undercooling. If the metastable phase diagram is known, one could readily incorporate this change in $C_0$ in the JH model. (c) The JH treatment has recently been extended by Magnin and Kurz [4] for irregular lamellar eutectics in which the interface deviates from planarity. At very low velocities, i.e. $V < 0.1$ μm/s in the Fe-C system, they found the constant $K_1$ to depend on the temperature gradient. However, for $V > 0.1$ μm/s, the effect of temperature gradient was found to be negligible and the results were nearly identical to the JH result, given by equation (4).

In summary, the basic result of the Jackson-Hunt model, given by equation (4), which relates the average interface undercooling with the eutectic spacing is valid up to growth rates of the order of ~ 10 mm/s if appropriate corrections for the variation in $C_0$ with undercooling are taken into account. Furthermore, the JH model also describes reasonably well both the regular and the irregular eutectics, the latter at velocities above ~ 0.1 μm/s. The basic JH model predicts multiple solutions for eutectic spacing at a given velocity, as shown in Fig. 5. We shall now discuss if there exists another physical criterion which selects a specific eutectic spacing at a given velocity.


II-2. Eutectic Spacing Selection

Based upon Zener's suggestion of the maximum growth rate hypothesis for eutectic growth in an undercooled melt [5], Tiller [6] proposed that, in a directional solidification experiment, the system will select the spacing which gives a minimum undercooling at the interface. This spacing is denoted by $\lambda_m$ in Fig. 5. Although the assumption of minimum undercooling gives a unique solution, there is no justification as to why a system would select this value. A more appropriate criterion should be based on the stability of the interface with respect to small changes in eutectic spacings*. JH qualitatively discussed this stability problem and concluded that all possible eutectic spacings for a given velocity, shown in Fig. 5, will not be stable against small fluctuations in spacings. They showed that eutectic spacings smaller than $\lambda_m$ will be unstable since any depression in the interface shape will cause a lamella at the depression to get narrower, which will ultimately eliminate that lamella and increase the local spacing, as shown in Fig. 6a. They also showed that the eutectic spacings larger than $\lambda_m$ will be stable. However, as the spacing becomes much larger than $\lambda_m$, the steady-state interface shape develops a pocket at the center of the wider phase, as shown in Fig. 6b. At some larger spacing, which we shall denote as $\lambda_M$, the slope of the interface becomes infinity so that all eutectic

---

*The word stable in this text refers to morphological stability with respect to fluctuation in the shape of the interface. If a fluctuation creates a change in the original growth form, the interface is called unstable.

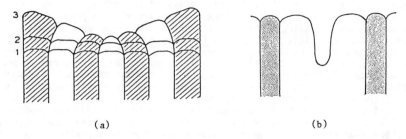

Fig. 6. (a) Schematic illustration of the instability of a lamellae with $\lambda < \lambda_m$. The lamellae at the center disappears with time. (b) The shape instability of the interface when the width of the phase becomes large [1]. A new lamella may be created in the depressed zone.

spacings above $\lambda_M$ become unstable. Thus, JH concluded that, of all possible spacings based on diffusional consideration for growth, only those spacings will be stable which lie in the band $\lambda_m < \lambda < \lambda_M$. The variations in $\lambda_m$ and $\lambda_M$ with velocity were calculated by JH. For the minimum stable spacing they obtained the relations:

$$V\lambda_m^2 = K_2/K_1, \text{ and} \tag{9}$$

$$\Delta T = [2 \sqrt{K_1 K_2}] V^{1/2}. \tag{10}$$

The relationship between $\lambda_M$ and V was obtained by assuming that a stable spacing, $\lambda > \lambda_m$, will exist until the slope of the interface for either of the phases becomes infinity. Using this criterion, they obtained the β-phase lamellar interface instability condition as

$$V \lambda^2_M = a(1 + b \sin \theta_\beta) (\Gamma D/m_\beta C_o), \tag{11}$$

where a and b are constants whose values depend on the volume fraction, $f_\beta$. The constant a was shown to vary from 325.8 to 368.5, and the constant b from 0.85 to 0.584, as $f_\beta$ was increased from 0.1 to 0.5.

The qualitative discussion of the stability of eutectic spacing gives two important conclusions. (1) For a given velocity, a band of eutectic spacings is stable, and the lower and upper limits of this stable spacing are given by $\lambda_m$ and $\lambda_M$. (2) Both, $\lambda_m$ and $\lambda_M$ satisfy the relationship $V\lambda_j^2$ = constant, j = m or M. The values of the constants are different for $\lambda_m$ and $\lambda_M$, so that if the physical constants for the system are known precisely, it would be possible to check experimentally whether the eutectic spacing exists randomly within the band or some selection process occurs which drives the spacings toward $\lambda_m$ or $\lambda_M$.

Since the work of Jackson and Hunt [1], several detailed stability analyses of eutectic interface have been carried out [7-14]. The treatments by Cline [7] and Hurle and Jakeman [8] considered only the average properties of eutectic interface so that the eutectic interface stability could be

11

treated as a single phase interface stability. A more realistic model was subsequently proposed by Cline [9] and Strassler and Schneider [10] who investigated the stability of a eutectic interface with respect to local variations in the spacing. The latter authors [10] concluded that the lamellar spacing smaller than $\lambda_m$ will be unstable, a result which was discussed qualitatively by Jackson and Hunt [1]. Furthermore, Strassler and Schneider found that the largest stable spacing, $\lambda_M$, depends on the temperature gradient (for off-eutectic alloys), and its value was found only numerically. Thus, the detailed stability analysis confirmed the prediction of JH that the stable eutectic spacing exists in the band $\lambda_m < \lambda < \lambda_M$. From the examination of the decay of perturbations, Strassler and Schneider concluded that the fastest decaying perturbation occurs when $\lambda$ is close to $\lambda_m$. They, thus, proposed that eutectic growth should always occur close to the minimum undercooling value.

The theoretical treatment by Strassler and Schneider assumed the interface to remain flat during perturbation, so that the motion of the triple point junction was not taken into account. This assumption was subsequently relaxed by Datye and Langer [13] who considered lamellar displacements parallel as well as perpendicular to the growth direction. They also concluded that $\lambda < \lambda_m$ is unstable. In addition, for sufficiently off-eutectic alloys, they predicted an oscillatory instability whose wavelength was twice the lamellar spacing.

The value of $\lambda_m$ is now well-established. However, we still do not have a satisfactory model for the $\lambda_M$ value. The Jackson-Hunt equation assumes the interface slope to go to infinity before the lamellae becomes unstable. It is quite possible that the spacing may become unstable when the slope of the interface in Fig. 6b reaches some finite value. Furthermore, it may be possible to nucleate the other phase in the pocket when the interface reaches some finite depth, thereby halving the local spacing. Consequently, the Jackson-Hunt criterion for $\lambda_M$ should be viewed only as an upper limit for $\lambda_M$.

A band of stable eutectic spacings, predicted for regular eutectic structures, also exists for irregular eutectics. Fisher and Kurz [15] have shown that a large variation in eutectic spacing can exist due to the stiffness in growth of the faceted phase. For the Fe-C system, the graphite flakes tend to grow in specific directions only so that the neighboring unaligned plates will either diverge or converge. The general model of irregular eutectic growth is shown in Fig. 7. When the spacing between the lamellae reaches the smallest stable value, the growth of that lamella stops. On the other hand, when the spacing between the lamellae reaches the largest stable spacing, a branch is created which will decrease the local spacing. Consequently, the eutectic spacing oscillates between $\lambda_m$ and $\lambda_M$, where $\lambda_M$ is the spacing at which branching occurs which is different from that given by equation (11). A detailed theoretical model for $\lambda_M$ was proposed by Magnin and Kurz [4], who showed that the branching occurs when the solid-liquid interface for the faceted phase develops a small depression whose base is slightly above the plane of the triple point junctions of that faceted lamella. They also showed that $\lambda_m$ can be characterized by the Jackson-Hunt result, given by equation (9). Since the lamellar spacing oscillates between $\lambda_m$ and $\lambda_M$, the average spacing of irregular eutectics will be larger than $\lambda_m$ by some factor $\phi$. Fisher and Kurz [15] proposed that the average spacing $\langle \lambda \rangle$ can be described as $\langle \lambda \rangle = \phi \lambda_m$.

The theoretical models for the regular and the irregular eutectics show that a finite band of stable eutectic spacings exists for a given velocity.

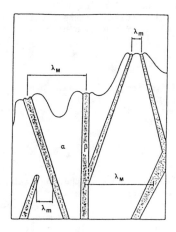

Fig. 7. Schematic diagram of the growth of an irregular eutectic showing the disappearance of a lamellae at $\lambda=\lambda_m$ and the branching of a lamellae at $\lambda=\lambda_M$.

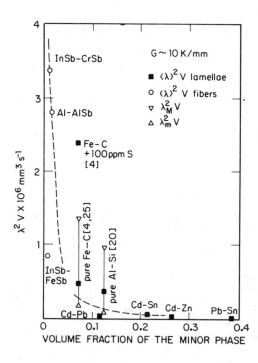

Fig. 8. The relationship between the experimentally measured $\langle K \rangle$ values and the volume fraction of the minor eutectic phase for systems with large $C_o$ values.

13

Thus, there appears to be no unique spacing selection, at least from the theoretical analyses that have been proposed so far. We shall, therefore, examine experimental results to answer the question of the uniqueness of the eutectic spacing under given growth rate condition.

## II-3. Experimental Observations

Directional solidification studies have been carried out in several binary alloy systems [16-29] to study the variation in average eutectic spacing, $\langle \lambda \rangle$, with velocity. These experimental results can be described as

$$V \langle \lambda \rangle^2 = \langle K \rangle, \qquad (12)$$

where $\langle K \rangle$ is the experimental value of the constant for a given system. Table 1 summarizes these experimental results. Two important parameters which influence the value of $\langle K \rangle$ are $f_\alpha$ and $C_o$. Figure 8 shows the variation in $\langle K \rangle$ with the volume fraction of the minor phase for systems with large $C_o$ values. Note that the $\langle K \rangle$ values for regular lamellar eutectics are of the same order of magnitude, whereas those for rod eutectics increase sharply as the volume fraction decreases, according to equation (9), as shown in Fig. 8. Furthermore, $\langle K \rangle$ values of irregular eutectics are significantly larger than those for the regular eutectics indicating that, for a given velocity, irregular eutectics will have coarser average spacings.

Table I.  Experimentally determined values of $\langle K \rangle$.

| Eutectic System $\alpha$-$\beta$ | Morphology | $\langle K \rangle$ mm$^3$/s | $f_\beta$ | Reference |
|---|---|---|---|---|
| Ag-Pb | Regular, Lamellar | $1.2 \times 10^{-7}$ | 0.85 | 16 |
| Ag$_3$Sn-Sn | Regular, Lamellar | $2.8 \times 10^{-7}$ | 0.97 | 16 |
| Al-Al$_2$Cu | Regular, Lamellar | $1.1 \times 10^{-7}$ | 0.46 | 16 |
| Al-Al$_3$Fe | Irregular, Lamellar | $9.2 \times 10^{-7}$ | 0.03 | 23 |
| Al-Al-Sb | Regular, Rod | $2.8 \times 10^{-6}$ | 0.01 | 21 |
| Al-Si | Irregular, Lamellar | $3.8 \times 10^{-7}$ | 0.13 | 20 |
| Al-Zn | Regular, Lamellar | $6.4 \times 10^{-8}$ | 0.70 | 16 |
| AuPb$_3$-Pb | Regular, Lamellar | $4.21 \times 10^{-8}$ | 0.52 | 18 |
| Bi-Zn | Regular, Lamellar | $6.9 \times 10^{-8}$ | 0.04 | 16 |
| Cd-Pb | Regular, Lamellar | $2.1 \times 10^{-8}$ | 0.81 | 16 |
| Cd-Sn | Regular, Lamellar | $7.2 \times 10^{-8}$ | 0.75 | 16 |
| Cd-Zn | Regular, Lamellar | $2.8 \times 10^{-8}$ | 0.17 | 16 |
| CrSb-InSb | Regular, Rod | $3.4 \times 10^{-6}$ | 0.99 | 22 |
| Fe-C | Irregular, Lamellar | $5.6 \times 10^{-6}$ | 0.07 | 4, 25 |
| Pd-Pb | Regular, Rod | $1.88 \times 10^{-8}$ | – | 18 |
| Pb-Sn | Regular, Lamellar | $3.3 \times 10^{-8}$ | 0.63 | 17 |
| Sn-Zn | Regular, Lamellar | $6.9 \times 10^{-8}$ | 0.09 | 16 |
| CAM-NAP* | Irregular, Lamellar | $2.1 \times 10^{-6}$ | – | 15 |
| CBr$_4$-C$_2$Cl$_6$ | Regular, Lamellar | $3.8 \times 10^{-7}$ | – | 19 |
| BOR-SCN* | Irregular, Lamellar | $1.5 \times 10^{-6}$ | – | 15 |
| NPG-SCN* | Irregular, Lamellar | $1.2 \times 10^{-6}$ | – | 15 |

*SCN = succinonitrile, BOR = borneol, CAM = camphor, NAP = Naphthalene, and NPG = neopentylglycol

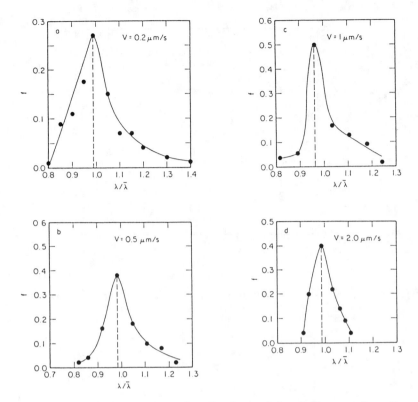

Fig. 9. Experimentally observed distributions of lamelallar spacings for different velocities in the $CBr_4$-$CCl_6$ system [19].

Jordan and Hunt [28] examined the experimental results on eutectic spacings in the Pb-Sn system and compared them with the JH model. They showed that the average spacings at different velocities were always larger than $\lambda_m$. Furthermore, they characterized the smallest and the largest spacings observed at a given velocity and showed that the deviation from the average spacing was significant. Similar results have recently been obtained by Seetharaman and Trivedi [19] in the transparent $CBr_4$-$C_2Cl_6$ system and by Trivedi, Mason and Kurz [18] in the Pb-Pd system. Figure 9 shows the distribution in lamellar eutectic spacing for different velocities in the $CBr_4$-$C_2Cl_6$ system. A significant variation in spacing is observed at a given velocity. Furthermore, because of the transparent nature of this system, it was possible to observe the disappearance of lamellae which had spacings less than $\lambda_m$. The distributions of spacings, observed at different velocities, are compared with the Jackson-Hunt model and this comparison is shown in Fig. 10. The average spacing is always about 20% larger than the minimum stable spacing. The variation in average spacing with velocity is shown in Fig. 11, in which the theoretical lines for $\lambda_m$ and $\lambda_M$ are also shown for comparison. Note that the average spacing also exhibits the same velocity dependence as the $\lambda_m$ or $\lambda_M$. Furthermore, the average spacings are closer to $\lambda_m$ than $\lambda_M$.

The experimental results for irregular eutectics also show a finite spread in the eutectic spacing at a given velocity. Figure 12 shows the

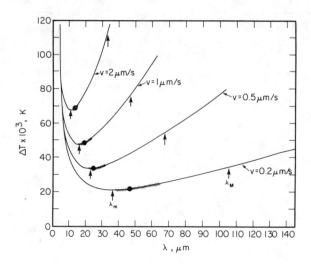

Fig. 10. A comparison of experimentally observed spacing distributions with the JM model for lamellar eutectic growth in the $CBr_4$-$C_2Cl_6$ system [19].

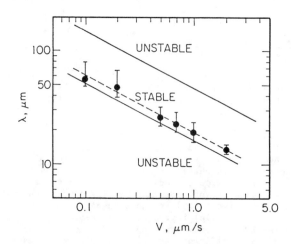

Fig. 11. The variation in $\langle\lambda\rangle$ with V in $CBr_4$-$C_2Cl_6$ system. The theoretical variations in $\lambda_m$ and $\lambda_M$ with V are also shown for comparison [19].

experimental data in the Fe–C system. The average spacing, in this case, is significantly larger than $\lambda_m$. In view of the experimental results for regular as well as irregular eutectic systems, we may in general describe the average experimental spacing as $\langle\lambda\rangle = \phi\,\lambda_m$, where $\phi$ is a constant for a given system.

The variation in the average eutectic spacing, $\langle\lambda\rangle$, and the average eutectic undercooling, $\langle\Delta T\rangle$, with velocity can now be generalized for both the regular as well as the irregular eutectic growth as follows [30].

16

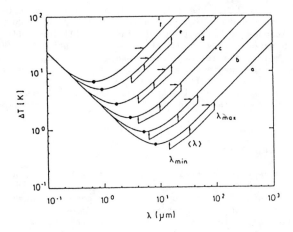

Fig. 12. A comparison between the experimentally observed spacing variation in an irregular Fe-C eutectic [25] and the theoretical model of JH.

$$V\langle\lambda\rangle^2 = \phi^2(K_2/K_1) \qquad (13)$$

$$\langle\Delta T\rangle\langle\lambda\rangle = (\phi^2+1)\,K_2, \text{ and} \qquad (14)$$

$$\langle\Delta T\rangle^2/V = [\phi + (1/\phi)]^2\,K_1\,K_2 \qquad (15)$$

The value of $\phi$ is about 1.2 for $CBr_4-C_2Cl_6$ regular eutectics, and it was found to be 2.5 for the pure Fe-C system [25] and 2.3 for pure Al-Si [20]. Equation (15) is the critical equation for determining the coupled zone for eutectic microstructures, and it is convenient to rewrite it in the following form

$$\langle\Delta T\rangle = K_E\,V^{1/2} \qquad (16)$$

where the constant $K_E$ is the square root of the right hand side of equation (15).

The above modification should allow one to design processing conditions to obtain a specific spacing. Note that the above results are valid only for velocities smaller than about 10 mm/s. Furthermore, for irregular eutectics the velocities should be larger than about 0.1 μm/s.

### III.  Eutectic Growth at High Velocities

The theoretical model of eutectic growth, discussed in the last section, predicts that the eutectic spacing is inversely proportional to the square root of velocity. Thus, by increasing the growth rate one may obtain an extremely fine eutectic spacing. However, the above relationship was

17

based on the assumption that the growth rates are small, so that this relationship may not be valid at high growth rates. The Jackson-Hunt model has recently been extended by Trivedi, Magnin and Kurz [2] for conditions of very high growth rates. They showed that (1) the relationship, $V\lambda^2$ = constant, does not hold at large velocities, and (2) the coupled eutectic growth will not occur above a certain critical velocity. These conclusions have been observed experimentally for eutectoid [31] and eutectic [32, 33] structures. We shall first outline the basic theoretical ideas which become important at high growth rates and then describe relevant experimental results.

III-1.  Theoretical Model

The major concepts which modify Jackson-Hunt's treatment at high velocities will be examined first.

(1)  The diffusion field is governed by the complete equation (2), and not by the Laplaces equation, as assumed in the JH-model.

(2)  At high velocities the undercoolings at the interface are large so that the variation in $C_0$ with velocity becomes important.

(3)  At large undercoolings the interface temperature decreases significantly so that the variation in the diffusion coefficient with temperature may become important. If $D = D_0 \exp(-Q/RT)$, where $D_0$ and Q are the pre-exponential constant and the activation energy, respectively, then $d\ln D/dT = Q/RT^2$. Thus, the variation in D becomes very significant as the value of the interface temperature T decreases. This effect will be important for systems with small solubilities and low eutectic temperatures.

(4)  At very high velocities, local equilibrium at the interface may not be established so that both the solute trapping effect [34, 35] and the proper non-equilibrium thermodynamic description of the interface conditions [36] should be taken into account. However, as we shall see later on, the cooperative eutectic growth cannot occur beyond a certain critical velocity, and this critical velocity is generally smaller than that at which solute trapping becomes important. Consequently, we shall not discuss this effect in this paper.

The diffusion field in liquid at high velocities is governed by the equation

$$(\partial^2 C/\partial X^2) + (\partial^2 C/\partial Z^2) + (V/D)(\partial C/\partial Z) = 0, \qquad (17)$$

for the lamellar eutectic growth. The solute diffusion process, described by equation (16), can be divided into two parts: (1) diffusion ahead of the interface, i. e. direction Z in Fig. 13, and (2) lateral diffusion from $\alpha$ to the $\beta$-phase, i. e. direction X in Fig. 13. The solute flux in directions Z and X will be inversely proportional to the diffusion distances, D/V and $\lambda/2$, respectively. At low velocities, D/V $\gg$ $\lambda/2$ so that the lateral flux predominates. Consequently, the last term on the left hand side of equation (17) is small and can be neglected. At large velocities, D/V becomes small so that it is of the same order of magnitude as $\lambda/2$ so that solute flux in both directions will be comparable. As the velocity becomes even higher, the solute transport occurs primarily in the direction Z, at which point the lateral diffusion becomes small so that the spacing will increase, and the interfacial energy will decrease, to compensate for the lower diffusional advantage in the lateral direction. The predominant diffusion in the Z direction at high velocities can also be seen from equation (17), in which the last term becomes more important at high velocities. When the velocity becomes sufficiently large that the first term in equation (17) becomes

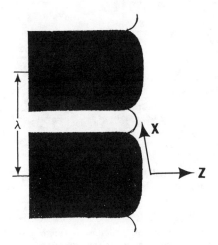

Fig. 13. Diffusion paths during the growth of a eutectic structure.

negligible compared to the last term, then no significant lateral diffusion occurs and eutectic structure will not be stable. Thus, we can conceptually understand why a cooperative eutectic growth will not occur above a certain critical velocity. This situation is equivalent to absolute stability in rapid single phase solidification.

A detailed solution of equation (16) was developed by Trivedi, Magnin and Kurz [2], and they showed that the result can be written as equivalent to the J-H result, given by equation (4), in which the parameter P is a function of the solute peclet number p, where $p = V\lambda/2D$. The function P (p) was shown to depend on the nature of the phase diagram. However, for the case, $k_\alpha = k_\beta = k$, they obtained

$$P\ (p) = \sum_{n=1}^{\infty} (\frac{1}{n\pi})^3 \ [Sin \ n\pi f_\alpha]^2 \ \frac{P_n}{\sqrt{1 + P_n^2} - 1 + 2k} \qquad (18)$$

where $P_n = n\pi/p$. Jackson and Hunt's [1] model predicted P to be a function of $f_\alpha$ only. However, at large growth rates, P depends on $f_\alpha$, p and k. The dependence of P on p and k is governed by the last factor in equation (18). At high velocities, p becomes large and $P_n$ becomes small, so that the last factor can be approximated as $2P_n/(P_n^2 + 4k)$. Thus, for finite k, P will approach zero at large velocities. For the special case, k = 0, the last factor varies inversely with $P_n$ so that P will go to infinity. This variation in the function P with p is shown in Fig. 14. In Fig. 14, the case k = 1 corresponds to the phase diagram with parallel solidus and liquidus lines.

We shall now examine the effect of high velocities on eutectic spacing. To the first order, equation (9) shows that $V\lambda_m^2$ is inversely proportional to $K_1$ and thus to P. Consequently, $V\lambda_m^2$ will increase sharply at high

19

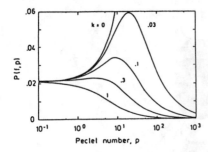 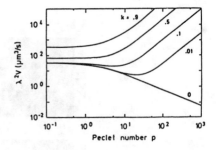

Fig. 14. The variation in the Function P with the peclet number.

Fig. 15. The variation in $\lambda_m^2 v$ with the peclet number for different k values.

velocities as P becomes small, or p becomes large. $\lambda_m$ becomes infinity at some critical velocity so that a cooperative eutectic growth cannot occur above this critical velocity. The variation in $V\lambda_m^2$ with the peclet number p is shown in Fig. 15 for different values of k and for $f_\alpha = 0.25$. Note that $V\lambda^2$ deviates sharply from the constant value at $p > 1$. The condition, $p = 1$, is simply that for which the diffusion distance ahead of the interface, $D/V$, is equal to the lateral diffusion distance, $\lambda/2$. Thus, the solution of the diffusion equation quantitatively shows the result that we had discussed qualitatively earlier.

The variation in $\lambda_m$ with velocity is shown in Fig. 16. For $k = 0.8$ the spacing will decrease less with velocity than that predicted by the JH model. Also the spacing will increase at high velocities and become infinity at some critical velocity. This limiting velocity can be understood if we examine the variation in undercooling with velocity. At the critical velocity the interface undercooling approaches the freezing range of one of the phases so that this phase will be preferentially selected over the eutectic structure. This limit of undercooling will be reached at lower velocities for systems with larger k (k near 1) values.

For systems with small k values, another effect can also become important which can cause the instability of the cooperative eutectic growth. This effect arises from the variation in the diffusion coefficient with the interface temperature. When k is small, large undercoolings can be achieved before the theoretical limit due to the large peclet number becomes critical. As the undercooling gets larger, the interface temperature decreases, so that the diffusion coefficient can decrease significantly with the increase in undercooling. As discussed earlier, the variation in D becomes more important for systems with very low eutectic temperatures. Since this effect must be considered simultaneously with the large peclet number effect, the actual behavior of spacing variation with velocity becomes quite complicated. However, in order to isolate this effect, we may consider a special case in which the variation in D with temperature is significant but the peclet number is still much less than one. For this case, we may use the JH result, given by equation (9), which predicts that $\lambda_m$ is directly proportional to the square root of D. Thus, a decrease in D will reduce the eutectic spacing. Since D decreases exponentially with temperature, a maximum in V vs. $\lambda$ will be observed [32], so that it will not be possible to obtain a cooperative eutectic growth beyond this maximum velocity.

20

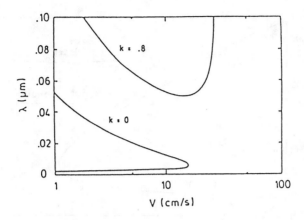

Fig. 16. The effect of high peclet number (k = 0.8 case) and of temperature
dependent diffusion coefficient (k = 0 case) on λ-V relationship.


Figure 16 illustrates the two important factors which give rise to the
limiting velocity for the cooperative eutectic growth. The case, k = 0.8,
illustrates the large peclet number effect, whereas the k = 0 case illustrates the diffusion coefficient effect. Note that, at high velocities, the
peclet number effect gives larger spacings and the D effect gives smaller
spacings compared to the JH result. Consequently, when both these effects
are important, the general variation in λ and V will show a more complex
behavior. For example, at high velocities, λ can decrease initially due to
the D effect but then increase due to the peclet number effect, or vice
versa [2].


III-2.  Experimental Observations

    Only limited experimental studies on eutectic growth have been carried
out at high velocities in which the eutectic interface velocity was measured
precisely. Boettinger et al. [32] used electron beam scanning to examine
the eutectic growth in the Ag-Cu system. They showed that a regular eutectic structure can be observed only up to a velocity of about 2.5 cm/s and at
this velocity the eutectic spacing was found to be 23.7 nm. They used
Jackson-Hunt's analysis with a temperature-dependent diffusion coefficient
to examine their results. The observed maximum velocity agreed well with
the theoretical prediction of 2.7 cm/s. However, the theoretical spacing of
10.6 nm was significantly smaller than the experimentally observed value of
23.7 nm.

    A detailed experimental study has recently been carried out by
Zimmerman et al. [33] in the Al-Cu system. They used a laser scanning technique and measured the variation in eutectic spacing with interface velocity, as shown in Fig. 17. They found that the JH model described the results for V ≤ 2.0 cm/s. From 2.0 to 20.0 cm/s, the data followed the prediction of Trivedi et al. [2] model quite well. Furthermore, the maximum
velocity for eutectic growth was shown to agree well with the theoretically
predicted value of 50 cm/s. However, between the velocities of 20.0 and 50
cm/s, two significant deviations were observed. First, the eutectic structure was not regular but was wavy in nature. Second, the eutectic spacing

21

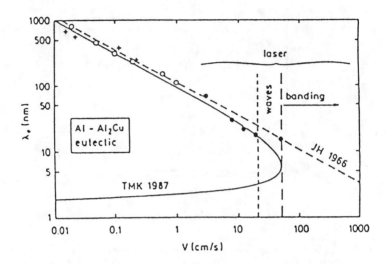

Fig. 17. Experimental results on the effect of velocity on eutectic spacing in the Al-Cu system. Theoretical predictions of the Jackson-Hunt and the Trivedi-Magnin-Kurz model are shown for comparison.

in this regime remained constant with a value of about 13 nm. It is indeed possible, as discussed previously, that first the temperature-dependent diffusion coefficient effect and then the high peclet number effect may become predominant to give this spacing variation behavior at high velocities. The reason for the wavy nature of the eutectic at high velocities is not yet well-understood. When the velocity was increased beyond 50 cm/s, no eutectic structure was observed. Instead, a banded structure, with bands perpendicular to the growth direction, was observed.

Another experimental result, which is germane to our discussion, is the directional transformation of austenite to pearlite which was studied by Pearson and Verhoeven [31] and by Bolling and Richman [37]. Pearson and Verhoeven [31] found that the cooperative growth was possible only up to a velocity of 106 μm/s. This value of the maximum velocity is consistent with the theoretical model since the diffusion coefficient of carbon in austenite is about $3.8 \times 10^{-8}$ cm$^2$/s, which is roughly three orders of magnitude smaller than the typical diffusion coefficient value in liquids. They found the pearlite spacing to be 55 nm at V = 106 μm/s. For velocities larger than 106 μms, a divorced pearlite was observed. A similar observation was made by Bolling and Richman [37] who also found that the spacing of the divorced pearlite didn't change as the velocity was increased. Pearson and Verhoeven [31] explained the constancy of spacing on the nucleation of pearlite colonies ahead of the interface. These pearlite colonies grew at the maximum possible rate, i.e. 106 μm/s, so that the spacing remained constant. The wavyness of the microstructure was attributed by them to the coarsening of very fine pearlite.

Experimental observations in eutectoid and eutectic growth appear to be quite similar. More, well-characterized, experimental studies in eutectoid and eutectic systems are, however, needed to see if the explanations presented for the eutectoid system are valid and if they could be applied to eutectic systems.

## IV. Microstructure and Phase Selection Diagrams

In our discussion of eutectic growth we have inherently assumed that a eutectic structure will form if proper diffusion coupling can be maintained between the two phases. Whether a eutectic structure would actually form under given conditions, however, depends on the stability of the eutectic interface with respect to a number of phenomena which we shall now discuss briefly.

(1) A given eutectic structure must be stable with respect to local fluctuations in spacings. This stability problem was discussed in section II-2 and it was concluded that a band of stable spacings exists for a given velocity.

(2) The eutectic interface should be stable with respect to periodic fluctuations in the macroscopic shape of the interface. Experimental studies show that the planar eutectic interface can become unstable and form eutectic cells or eutectic dendrites. Such instabilities form when a third element is present which is rejected by both phases. Since the final microstructure still exhibits a eutectic structure, we will not further discuss this stability problem.

(3) The eutectic structure should also be stable with respect to fluctuations which lead to a single phase planar, cellular or dendritic interface. Such an instability would occur if one of the eutectic phases grows out from the eutectic interface and forms a planar single phase front or a cellular (or dendritic) structure in which a eutectic microstructure is present in the intercellular (or interdendritic) regions.

(4) The given eutectic structure should be stable with respect to the formation of other metastable phases. At high growth rates, metastable single phases or another metastable eutectic may be preferred.

The last two types of eutectic stability considerations are important in determining the conditions under which a eutectic microstructure will be present. Here, we would consider a more general problem which addresses the microstructural selection criterion for both the stable as well as metastable phases. The results of such microstructural and phase selection criterion can then be superimposed on the general stable and metastable phase diagram to visually map out the conditions which give rise to all possible microstructures in a given system.

The most desirable approach to develop a microstructural selection criterion is to carry out a stability analysis of the interface. However, because of the complexity of the eutectic structure, it is expected that nonlinear effects would be critical in the microstructural selection process. Consequently, we shall use a simpler approach which considers all possible growth morphologies under given experimental conditions and then assumes that the observed morphology will be the one that leads other morphologies during the growth. Since the leading morphology will have the highest temperature, or the lowest undercooling, one can determine the stable morphology by comparing interface temperatures of planar, cellular, dendritic and eutectic microstructures. This approach was developed quantitatively by Hunt and Jackson [45, 46] and later by Burden and Hunt [38] to determine the range of velocities over which a eutectic or dendritic structure will exist. Pursuing an idea by Fredriksson [47], Kurz and Fisher [39] extended this approach to irregular eutectics and to plate dendrites of β which are the common faceted phase morphologies in irregular eutectic systems and they calculated coupled growth zones for several alloy systems.

23

All these approaches used dendrite and eutectic models which are not valid under rapid solidification conditions. We shall therefore examine the general models [2, 40] which predict dendrite and eutectic interface temperatures under low as well as high velocity conditions. We shall first examine the theoretical ideas for the microstructure selection process and then extend these ideas to the metastable phase formation.

IV-1. Interface Temperatures

The variation in the eutectic interface temperature, $T_i(E)$ with undercooling is given by Trivedi et al. [2] as

$$T_i(E) = T_E - [2\sqrt{K_1'} \, K_2] \, (P/D)^{1/2} \, [(1 + F/2) \, / \, (1 + F)^{1/2}] \, V^{1/2} \qquad (19)$$

where $K_1' = mC_0/f_\alpha f_\beta$ and $F = d\ln P/d\ln\lambda$. From the table of values, given by Trivedi et al. [2], the value of the term $[(1 + F/2) \, / \, (1 + F)^{1/2}]$ is close to unity for conditions which give stable cooperative eutectic growth. The major effects at high growth rates come from the velocity dependent P and the temperature dependent D. For small velocities, P and D are constant and equation (19) simplifies to equation (10).

A general model of columnar dendritic growth, which is valid at high velocities, is developed by Kurz et al. [40]. We shall modify their results to include the low velocity cellular growth regime in which the term GD/V dominates the interface undercooling [38]. The interface temperature, $T_i$ ($\alpha$), for the single phase $\alpha$, can be described by

$$T_i(\alpha) = T_L - (GD/V) - [k\Delta T_0 \, Iv(p) \, A] - \Gamma \, \kappa_t, \qquad (20)$$

where $T_L$ is the liquidus temperature of the alloy, $\Delta T_0$ is the freezing range of the metastable single phase, p is the solute peclet number equal to VR/2D in which R is the dendrite or cell tip radius and $\kappa_t$ is the curvature of the dendrite tip which is equal to 1/R for plate dendrites and 2/R for needle dendrites. $Iv(p)$ is the Ivantsov function given as

$$Iv(p) = \begin{cases} p \, \exp \, (p) \, E_1(p) & \text{for needle dendrite} \\ \sqrt{\pi p} \, \exp \, (p) \, \text{erfc} \, (\sqrt{p}) & \text{for plate dendrite.} \end{cases} \qquad (21)$$

The parameter A is the ratio of tip to initial alloy concentration, and it is given by the relationship:

$$A = [1-(1-k) \, Iv(p)]^{-1} \qquad (21a)$$

The cell or dendrite tip radius is given by

$$R = [\Gamma/2p\Delta T_0 \sigma^*] \, [Ak\zeta_c - (GD/V\Delta T_0)]^{-1} \qquad (22)$$

in which $\sigma^*$ is the stability constant and the function $\zeta_c$ is given by

24

$$\zeta_c = 1 - \frac{2k}{[1 + (1/\sigma^* p^2)]^{1/2} - 1 + 2k} \qquad (23)$$

At high velocities the solute distribution coefficient will be a function of velocity and the relationship between k and V is obtained by Aziz [34] as

$$k = [k_o + (a_o V/D)]/[1 + (a_o V/D)], \qquad (24)$$

in which $k_o$ is the equilibrium solute distribution coefficient and $a_o$ is a length scale which is of the order of the interatomic distance.

The variation in the tip temperature of the single phase is given by equation (20). Since the general equation is quite complex, we shall illustrate the result in Fig. 18. The regions of planar, cellular and dendritic interfaces are indicated in the figure.


IV-2. Microstructural Selection

A comparison of equations (19) and (20) would give the velocity range, for a given composition, where eutectic or single phase microstructure with eutectic will be present. A eutectic microstructure will exist when $T_i(E)$ > $T_i(\alpha)$, and a single α-phase with interdendritic eutectic will be present when $T_i(\alpha) > T_i(E)$. Note that, in general, both α as well as β single phase interface temperatures should be considered. We shall, however, first discuss the case where α and β single phases have similar properties so that only one of the phases is stable for a given composition. A more general case will be considered later on.

Figure 19 shows schematically the variation in the eutectic and the single phase interface temperatures with velocity for hypereutectic composition. For this eutectic system we have assumed that k is small so that the diffusion coefficient effect becomes predominant at high velocities which gives rise to a maximum in the velocity. Figure 19 shows that a eutectic structure will be present at very low velocities and also at some higher velocities. The various microstructures which will exist at different velocities are also indicated in the figure. Furthermore, the interface temperatures at which these transitions occur are also shown on the right hand side of the figure. The undercoolings for different microstructures correspond to the eutectic undercooling equation (19) until the maximum undercooling for the eutectic growth is obtained. At higher velocities, the single phase undercooling equation (20) will give dendrite to cellular transition. A planar interface transition at high velocities occurs at the velocity for absolute stability of a planar interface as predicted by Mullins and Sekerka [41], where the interface temperature equals the solidus temperature of the metastable single phase.

The changes in microstructures with undercooling, shown in Fig. 19 for a given composition, can be superposed on the phase diagram. If such calculations are carried out for different compositions, one can determine the zones of interface temperature and composition where various microstructures will be formed.


IV-3. The Coupled Zone

Before we discuss the composition dependence of various transition points, we shall first examine the special case of the eutectic microstruc-

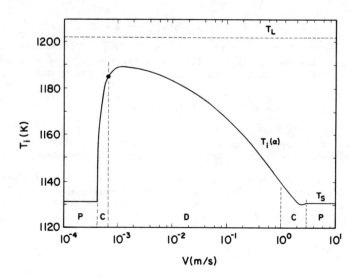

Fig. 18. A schematic variation in the single phase interface temperature with velocity and its correlation with planar, cellular and dendritic structures for $G > 0$.

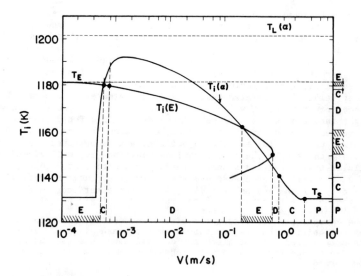

Fig. 19. The variation in eutectic temperature with velocity is compared with that for the single phase. Relevant microstructures as a function of velocity or undercooling are also shown.

ture zone. Let $T^*$ and $C^*$ define the boundary between the α-phase (with interdendritic eutectic) and the completely eutectic zones of Fig. 1. In order to understand the general characteristics of the $(T^*, C^*)$ line, we will first examine the coupled zone for the two limiting cases; the low and the high undercooling conditions. From the results of these limiting cases, we shall discuss why some systems exhibit regular coupled zones whereas the others exhibit skewed coupled zones. The parameters which play a key part in determining the regular or the skewed nature of the coupled zone will then be identified. Note that the widening of the coupled zone at low undercoolings is observed only in directional solidification where a positive temperature gradient stabilizes eutectic microstructures. For undercooled alloy melt, the coupled zone simply approaches the eutectic composition when $T^* \to T_E$.

At low velocities, the eutectic interface temperature is given by equation (16) as

$$T_i(E) = T_E - K_E V^{1/2} .$$
(25)

From Fig. 19 it is seen that the eutectic structure becomes stable at low velocities due to the sharp drop in the cellular/dendritic interface temperature. This sharp change in the interface temperature occurs due to the presence of the positive gradient through the term, GD/V in equation (20). Thus, for finite G, the single phase interface temperature is given by equation (20) in which the contributions from the second and the third term on the right hand side are negligible.

$$T_i(\alpha) \simeq T_L - GD/V .$$
(26)

By equating equations (25) and (26), we obtain the value of the velocity at which the dendrite - eutectic transition occurs at low undercoolings. Substituting this value of velocity in equation (25) gives the equation of the high temperature part of the coupled zone boundary, $(T^*, C^*)$, as

$$-m_\alpha(C^* - C_E) = GDK_E^2/[T_E - T^*]^2 - [T_E - T^*] .$$
(27)

The above equation predicts that $T^*$ decreases as $C^*$ is increased. The slope of this line is approximately given by

$$(dT^*/dC^*) = -m_\alpha/[1 + (2GDK_E^2)/(T_E - T^*)^3] .$$
(28)

Since the term in the large bracket on the right hand side of equation (28) is always positive, and $m_\alpha$ was defined as positive, the coupled zone boundary between the α and the eutectic phase will have a negative slope which increases with the increase in undercooling (or composition). The result of the low undercooling coupled zone, given by equation (27), is shown in Fig. 20. Note that the location of this line depends on the temperature gradient imposed upon the system, and for this reason it is indicated by a dotted line in Fig. 20.

For the high undercooling regime, the actual interface temperature equations are more complex. Thus, in order to understand the general form

27

of the coupled zone, we shall assume that the dendrite-eutectic transition velocities are in the regime where the peclet numbers are of the order of unity or less. Under this condition, the eutectic interface temperature can be approximated by equation (25). At high undercoolings, for the dendrite tip temperature, the terms GD/V and $\kappa_t$ in equation (19) are small and can be neglected. Kurz and Fisher [39] have shown that under this assumption, equation (19) can be approximated for needle dendrite as

$$T_i(\alpha) = T_L - K_N C^{1/2} V^{1/2} \, ,$$

where
$$K_N = A_N [m_\alpha (1-k_\alpha) \Gamma/D]^{1/2} \, . \tag{29}$$

Equating equation (25) and (29) gives the velocity at which the dendrite-eutectic transition occurs at higher growth rates. Substituting the value of this velocity in equation (25), we obtain the relationship between $T^*$ and $C^*$ as

$$m_\alpha (C_E - C^*) = [K_R C^{*1/2} - 1][T_E - T^*] \, , \tag{30}$$

where $K_R = K_N/K_E$. The slope of the coupled zone line at high undercoolings can then be approximated as[+]

$$(dT^*/dC^*) = [m_\alpha + (K_R/2\sqrt{C^*})(T_E - T^*)]/[K_R \sqrt{C^*} - 1] \, . \tag{31}$$

Equation (30) and (31) predict two distinctly different behaviors of the coupled zone boundary at high undercooling values, and these behaviors depend on the sign of the factor $K_R C^{*1/2} - 1$. We shall now consider these two cases separately.

Case I: $K_R C^{*1/2} > 1$.

In this case, the right hand side of equation (30) is positive so that real solutions are obtained only for compositions $C^* < C_E$. Also the slope of the coupled zone line, given by equation (31), is positive. This case is denoted in Fig. 20 as I. The curve I, when combined with the low undercooling curve, will give rise to the coupled zone boundary between the $\alpha$-phase and the eutectic phase, as shown in Fig. 1a. If the $\beta$-phase also exhibits a similar behavior, then a regular coupled zone would exist in the system.

We shall now examine the condition which gives rise to the curve I. Since $K_R$ is a constant, the condition $K_R C^{*1/2} > 1$ is satisfied for compositions $C^* > (1/K_R^2)$ or $C^* > C_{crit}$, where $C_{crit} = 1/K_R^2$. Since $C_{crit}$ is the

---

[+]The constant $K_E$ has a small composition dependence through the volume fractions of eutectic phases. This effect is small for lamellar eutectics and is ignored in calculating the slopes.

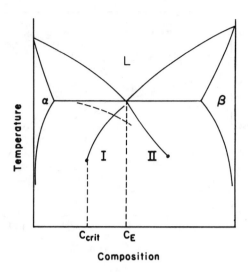

Fig. 20. The low and the high under cooling branches of the coupled zone
boundaries between the $\alpha$-phase and the eutectic.

property of the system[+], case I will exist only when $C_{crit} < C_E$. Under this
condition, the high undercooling branch of the coupled zone would exist only
for compositions, $C_{crit} < C^* < C_E$ .

Case II: $K_R C^{*1/2} < 1$.

For this condition, real solutions of equation (30) are obtained only
for $C^* > C_E$. The slope of the coupled zone is predicted by equation (31) to
be negative. Thus, the high undercooling branch of the coupled zone will
appear as line II in Fig. 20. This result, when combined with the low
undercooling branch, will give rise to a skewed coupled zone, as shown in
Fig. 1b. In Fig. 1b, the $\beta$-phase is shown to follow the case I. Note that
for the $\beta$-phase the concentration axis is reversed. If the $\alpha$-phase follows
the case I and the $\beta$-phase follows the case II, then the coupled zone in
Fig. 1b will be skewed to the left.

The existence of a regular or a skewed coupled zone depends on the
value of $C_{crit}$, which is the property of the system. If $C_{crit} < C_E$, a
regular zone is predicted, whereas $C_{crit} > C_E$ gives rise to a skewed coupled
zone. Since $C_{crit}$ is inversely proportional to the square of the constant
$K_R$, a regular coupled zone would require larger $K_R$ for both phases, i.e.
larger $K_N$ and smaller $K_E$. A skewed coupled zone, on the other hand, would
require very small $K_R$ for one of the phases, i.e. smaller $K_N$ and larger $K_E$.

---

[+]For more precise calculations of $C_{crit}$, the effect of composition on volume
fractions in the parameter $K_E$ should be taken into account.

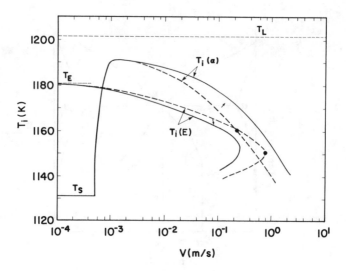

Fig. 21. The shifts in dendrite tip or eutectic interface temperature required to eliminate the high velocity region of eutectic growth at
$C < C_E$ and produce a skewed coupled zone.

The above discussion was based on mathematical equations for the coupled zone boundary. We may obtain a better insight into the conditions for a
regular or a skewed coupled zone by examining the relative interface
temperatures of the α-phase and the eutectic phase interfaces, as illustrated in Fig. 21. If the eutectic and dendrite temperatures follow the behavior indicated by dotted lines in Fig. 21, a eutectic zone will be observed
at high velocities. However, this zone will be absent if the dendrite tip
temperature is given by the solid line or if the eutectic temperature is
given by the solid line. These shifts in temperatures will occur if $K_N$ is
smaller or $K_E$ is larger, i.e. if $K_R$ is smaller. Thus, a skewed zone would
be predicted for systems with large $K_E$ values, as in the cases of irregular Al-Si and Fe-C eutectics, and/or very small $K_N$ value of the α phase. It
should be emphasized that the absence of high undercooling zone will occur
when the high peclet number contribution or the temperature dependent
diffusion coefficient effect is taken into account for the eutectic growth,
the latter effect is quite clear in Fig. 21.

The skewed zone boundary between the α and the eutectic phase, for $C^* \leq$
$C_E$, is governed by the low undercooling branch only so that its shape is
given by equation (27). In such a system, a eutectic composition will exhibit a eutectic structure only for undercoolings, $\Delta T^*(C_E)$, such that

$$\Delta T^*(C_E) \leq (GDK_E^2)^{1/3} \tag{32}$$

The extent of the skewed coupled zone also depends on the coupled zone
boundary between the β-phase and the eutectic. For example, the width of
the zone would be increased if the high undercooling branch of the coupled
zone boundary for the β-phase is shifted upwards. This would require larger

30

$K_N$ values for the β-phase. It was shown that a larger value of this constant would be obtained if the β-phase formed as a plate rather than a needle [42], which is indeed observed for primary C or Si in the corresponding systems [39]. In this case, the dendrite tip temperature equation is modified as

$$T_i(\beta) = T_L - K_p C^{1/n} V^{1/n}$$  (33)

The values of $n \simeq 3$ and $K_p C^{1/n} = 930$ Ks$^{0.33}$ mm$^{-0.33}$ were found for primary carbon plate in the Fe-C system. This larger value of $K_p$, compared to $K_N$, will increase the $K_R$ value for the β-phase, and thus will increase the coupled zone width significantly. In fact, if the value of $K_N$ or $K_p$ for the β-phase is not large, there may not be a significant coupled zone.

## VI-4. Microstructure and Phase Selection

In Figs. (18) and (19), the changes in microstructures with velocity or undercooling were illustrated for a given composition. If similar calculations are carried out for different compositions, then one may exhibit the temperature – composition zones for various microstructures in a manner similar to that carried out for the eutectic zone. Figure 22 is a schematic drawing of the various microstructural features that would be observed as a function of interface temperature. We have included the banded structure which is observed experimentally, at least in Ag-Cu [32] and Al-Cu [33] systems, when the eutectic structure can no longer be formed. Also shown schematically is the glass transition temperature, $T_g$, which is important for eutectic systems since eutectic systems have been shown to form glass readily compared to non-eutectic systems.

The microstructure diagram, shown in Fig. 22, can also be extended to include stable as well as metastable phases if we assume that the nucleation of metastable phase occurs readily [43]. Since the ideas are similar, we shall only illustrate this point schematically by using the metastable phase diagram for the Fe-Al system.

The phase diagram, Fig. 23, shows that when alloys of different compositions are directionally solidified, various phases can form at different undercoolings. These phases and their microstructures are: $Al_3Fe$ plate dendrites ($\beta_p$), $Al_6Fe$ needle dendrites ($\gamma_N$), $Al-Al_3Fe$ eutectic ($E_{\alpha\beta}$), $Al-Al_6Fe$ eutectic ($E_{\alpha\gamma}$) and Al needle dendrites ($\alpha_N$). There is at least one other metastable phase in this system. As its composition and its position in the metastable phase diagram have not been determined, it will not be discussed here. Figure 23 shows the theoretically calculated microstructure diagram for the Al-Fe system [44]. This diagram exhibits the temperature – composition regions of both the stable and the metastable phase microstructures, and it matches the diagram determined experimentally by Hughes and Jones [48]. Similar microstructural diagrams in velocity – composition coordinates have been discussed by Boettinger [49].

## V. Conclusion

The current status of the theoretical modeling of eutectic microstructures is critically examined and the progress made recently in this area is outlined. The Jackson-Hunt model is shown to describe remarkably well both the regular and the irregular eutectic growth at low velocities. However, there is sufficient experimental evidence to believe that the operating

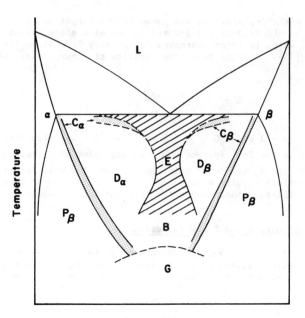

**Composition**

Fig. 22. The interface temperature and alloy composition regions illustrating various microstructural zones. P = planar, C = cellular, D = dendritic, E = eutectic, B = banded and G = glassy structures.

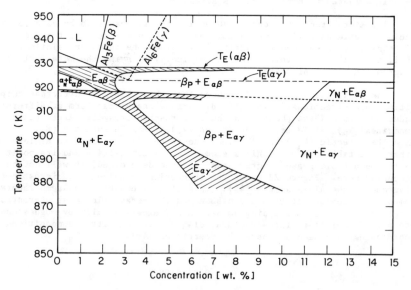

Fig. 23. The theoretical microstructure map for the Al–Fe system. For eutectic microstructures, $\phi = 1$ and $\phi = 3$ were assumed for the $\alpha$–$\gamma$ and $\alpha$–$\beta$ eutectics, respectively.

32

point of eutectic spacing is not at the minimum undercooling but slightly higher for regular eutectics and significantly higher for irregular eutectics.

A significant departure from the Jackson–Hunt relationship occurs when the velocity is very high. Furthermore, the eutectic structure is shown to be unstable above some critical velocity. This instability exists because of the high peclet number effect on the diffusion in liquid or because of the sharp decrease in the diffusion coefficient at low interface temperatures.

A detailed model for the prediction of coupled zones is developed and the relative stability of planar, cellular, dendritic and eutectic interfaces is discussed. The microstructural transitions are superposed on the phase diagram leading to microstructural diagrams which give a map of the correlation between microstructures and processing conditions.

## Acknowledgments

This work was carried out in part at the Ames Laboratory which is operated for the U. S. Department of Energy by Iowa State University under contract no. W-7405-ENG-82. This work was supported by the Office of Basic Energy Sciences, Division of Materials Sciences. One of the authors (RT) would also like to acknowledge the Alcoa Foundation for a grant which supported part of this work.

## References

1. K. A. Jackson and J. D. Hunt, Metall. Trans. $\underline{236}$, 1129 (1966).
2. R. Trivedi, P. Magnin and W. Kurz, Acta Metall. $\underline{35}$, 971 (1987).
3. R. W. Series, J. D. Hunt and K. A. Jackson, J. Cryst. Growth $\underline{40}$, 221 (1977).
4. P. Magnin and W. Kurz, Acta Metall. $\underline{35}$, 1119 (1987).
5. C. Zener, Trans. Metall. Soc. $\underline{167}$, 550 (1946).
6. W. A. Tiller, "Liquid Metals and Solidification", ASM, Cleveland, OH, p. 276 (1958).
7. H. E. Cline, Trans. Metall. Soc. $\underline{242}$, 1613 (1968).
8. D. T. J. Hurle and E. Jakeman, J. Cryst. Growth $\underline{3}$, 574 (1968).
9. H. R. Cline, J. Appl. Phys. $\underline{50}$, 4780 (1979).
10. S. Strassler, and W. R. Schneider, Phys. Cond. Matter $\underline{17}$, 153 (1974).
11. G. E. Nash, J. Cryst. Growth $\underline{38}$, 155 (1977).
12. J. S. Langer, Phys. Rev. Lett. $\underline{44}$, 1023 (1980).
13. V. Datye and J. S. Langer, Phys. Rev. $\underline{B24}$, 4155 (1981).
14. V. Datye, R. Mathur and J. S. Langer, J. Stat. Phys. $\underline{29}$, 1 (1982).
15. S. J. Fisher and W. Kurz, Acta Metall. $\underline{28}$, 777 (1980).
16. A. Moore and R. Elliott, "The Solidification of Metals", The Iron and Steel Inst., London, p. 167 (1968).
17. J. D. Hunt and J. P. Chilton, J. Inst. Met. $\underline{92}$, 21 (1963).
18. R. Trivedi, J. T. Mason and W. Kurz, Acta Metall. (to be published).
19. V. Seetharaman and R. Trivedi, Metall. Trans. (to be published).
20. R. G. Grugel and W. Kurz, Metall. Trans. $\underline{18A}$, 1137, (1987).
21. G. Beghi, G. Piatti and K. N. Street, J. Mater. Sci. $\underline{6}$, 118 (1971).
22. A. Muller and M. Wilhelm, J. Phys. Chem. Sol $\underline{26}$, 2029 (1965).
23. A. J. McLeod, L. M. Hogan, C. Adam and D. C. Jenkinson, J. Cryst. Growth $\underline{19}$, 301 (1973).
24. B. Toloui and A. Hellawell, Acta Metall. $\underline{24}$, 565 (1976).
25. H. Jones and W. Kurz, Z. Metallk. $\underline{72}$, 792 (1981).

26. J. N. Clark and R. Elliott, Met. Sci. J. $\underline{10}$, 101 (1976).
27. H. E. Cline, Mat. Sci. Engr. $\underline{65}$, 93 (1984).
28. R. M. Jordan and J. D. Hunt, Metall. Trans. $\underline{3}$, 1385 (1972).
29. H. E. Cline and J. D. Livingston, Trans. Met. Soc. $\underline{245}$, 1987 (1969).
30. W. Kurz and R. Trivedi, Proc. of Int. Conf. on Materials Processing, Sheffield, England (1987).
31. D. D. Pearson and J. D. Verhoeven, Metall. Trans. $\underline{15A}$, 1037 (1984).
32. W. J. Boettinger, D. Shechtman, R. J. Schaefer and F. S. Biancaniello, Metall. Trans. $\underline{15A}$, 55 (1984).
33. M. Zimmerman, M. Carrard and W. Kurz, Unpublished work, Swiss Federal Institute of Technology, Lausanne, Switzerland (1987).
34. M. J. Aziz, J. Appl. Phys. $\underline{53}$, 1158 (1982).
35. R. Trivedi, J. Lipton and W. Kurz, Acta Metall. $\underline{35}$, 965 (1987).
36. W. J. Boettinger and S. R. Coriell, in Science and Technology of Undercooled Melt, Ed. by P. R. Sahm et. al., Martinus Nijhoff, Utrecht (1986).
37. G. F. Bolling and R. H. Richman, Metall. Trans. $\underline{1}$, 2095 (1970).
38. M. H. Burden and J. D. Hunt, J. Cryst. Growth $\underline{22}$, 368 (1974).
39. W. Kurz and D. J. Fisher, Int. Met. Rev., 177 (1979).
40. W. Kurz, B. Giovanola and R. Trivedi, Acta Metall. $\underline{34}$, 823 (1986).
41. W. W. Mullins and R. F. Sekerka, J. Appl. Phys. $\underline{35}$, 444 (1964).
42. W. Kurz, Z. Metallk. $\underline{69}$, 433 (1978).
43. H. Jones and W. Kurz, Metall. Trans. $\underline{11A}$. 1265 (1980).
44. M. Gremaud, W. Kurz and R. Trivedi, Unpublished work, (1987).
45. J. D. Hunt and K. A. Jackson, Trans. AIME $\underline{239}$, 869 (1967).
46. K. A. Jackson, Trans. AIME $\underline{242}$, 1275 (1968).
47. H. Fredricksson, Metall. Trans. $\underline{6A}$, 1658 (1975).
48. I. R. Hughes and H. Jones, J. Mat. Sci. $\underline{11}$, 1781 (1976).
49. W. J. Boettinger, in Rapidly Solidified Amorphous and Crystalline Alloys, (B. H. Kear and B. C. Giessen, Eds) Elsevier, New York (1982).

# COMPUTER MODELING OF EUTECTIC GROWTH

Alain Karma

Condensed Matter Physics 114-36
California Institute of Technology
Pasadena, CA 91125

We present a numerical algorithm which can simulate the dynamics of interfacial pattern formation during eutectic growth. The algorithm uses random-walkers on a two-dimensional lattice which mimic molecular diffusion. The rules governing the escape and attachments of these walkers to and from the solid-liquid interface contains the physics of the crystal growth and surface tensions between different phases. We apply this algorithm to study the morphological stability of thin lamellar eutectics at off-eutectic melt compositions. Two unstable modes are observed in simulations. An oscillatory mode on twice the lamellar spacing and a tilting mode on the lamellar spacing. A simple theoretical interpretation of both instabilities is included and simulated eutectic morphologies are compared with experimental observations.

Solidification Processing of Eutectic Alloys
D.M. Stefanescu, G.J. Abbaschian and R.J. Bayuzick
The Metallurgical Society, 1988

# Introduction

The pioneering analysis of Jackson and Hunt (1) has provided the theoretical basis for our understanding of both lamellar and rod eutectic growth. Their analysis applies to spatially periodic eutectic structures forming under steady-state growth condition. During steady-state growth, the shape of the solid-liquid interface is preserved and it is essentially this time independence of the interface shape which renders steady-state growth amenable to analytical treatment.

However, in many important situations, the solid-liquid interface constantly changes its shape during the growth process. Non-stationary interface shapes are characteristic of systems where facets are present on the solid-liquid interface of one solid phase ($f - nf$ systems), but also occur frequently in systems where the solid-liquid interfaces of both solid phases are non-faceted (we shall only consider in what follows this second class of $nf - nf$ systems). For example, in three dimensions, rearrangements of lamellar structures are made possible by the motion of lamellar faults. These faults are likely to play a role in selecting the operating point of the system (i.e., the value of the lamellar spacing) and permit an interplay between lamellar and rod morphologies. Understanding their motion is of primary importance and requires a time-dependent description of eutectic growth. Also, in practice, steady-state growth is limited to a small range of melt compositions in the vicinity of the eutectic composition. Outside this range the solid-liquid interface becomes time-dependent and a wide variety of eutectic morphologies can form. Our knowledge of these morphologies has remained very limited. On the experimental side, the best documented non-steady-state behavior is one in which dendrites of one solid phase and lamellar eutectics both coexist (2). This situation is observed at off-eutectic compositions when the growth of one solid phase becomes sufficiently enhanced to cause dendrites of this phase to emerge ahead of the eutectic interface. There is also experimental evidence for dynamical modes of growth in which dendrites are absent (3,4). One of them is an oscillatory mode on twice the lamellar spacing which causes lamellae from the minor phase (the phase of smaller volume fraction) to follow sinusoidal solidification paths, as opposed to the vertical paths followed during steady-state growth. An experimental photograph of this mode is included in this paper (3). Other modes, which seem to involve more convoluted solidification paths of these lamellae, generate complex eutectic patterns ranging from orderly to chaotic.

An important tool which is missing at this point to model this large class of time-dependent growth problems is a numerical algorithm capable of simulating the motion of a eutectic solidification front (i.e., a tool to model non-steady-state growth). To construct such an algorithm, even for the simplest two dimensional situation characteristic of thin films, represents a formidable task since one must add the constraint of mechanical equilibrium at solid-solid-liquid triple points (where three phases meet in space) to the already existing difficulties of the free-boundary problem for a single solid phase.

In this paper I shall discuss a numerical algorithm which can simulate eutectic growth of $nf - nf$ systems (5). An example of a simulated lamellar structure is shown in Fig. 1. This algorithm is based on a random walk model originally developed by Kadanoff (6), Tang (7), and Liang (8) to study a hydrodynamics problem (viscous flow in a Hele-Shaw cell). Here, in the context of eutectic growth, random-walkers on a lattice mimic molecular diffusion between neighboring solid phases. The rules governing the escape of walkers from the interface contain the physics of the crystal growth and are constructed in such a way to preserve the balance of surface tensions at triple points. The main virtue of this algorithm is its ability to simulate almost arbitrarily complex growth processes and to help visualize the interfacial deformations underlying the formation of eutectic microstructures.

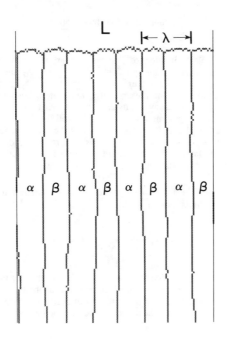

Fig. 1          Simulated lamellar eutectic morphology.

Next, I shall present some results of simulation which pertain to the morphological stability of lamellar eutectics at off-eutectic melt compositions. On the theoretical side, a complete stability analysis of lamellar eutectics has never been achieved, mainly because the coupling between interface displacements and the motion of solid-solid-liquid triple points complicate the problem enormously. Hurle and Jakeman have treated the eutectic front as a single phase and have predicted that lamellar eutectics should become unstable when the wavelength of the Mullins-Sekerka instability, induced by the boundary-layer of excess concentration of one of the component, becomes comparable with the lamellar spacing (9). This criterion undoubtedly captures the essential physical mechanism causing instability but can not predict the precise interface deformations associated with this type of instability. Dayte and Langer have performed a more detailed morphological stability analysis of thin lamellar eutectics (10). Their analysis is based on constructing a discrete model for displacements of triple points by averaging the composition field over individual lamellae and predicts successfully the existence of an oscillatory instability on twice the lamellar spacing at sufficiently off-eutectic melt compositions.

We have observed in simulations that two instabilities are present at off-eutectic melt composition. The oscillatory instability (Fig. 3b) predicted by Dayte and Langer and a secondary tilting instability which occurs at a more off-eutectic melt composition (or larger growth rate). The interface deformation associated with the latter is shown in Fig. 3a.

37

## Equations of Interface Motion

We consider the simplest situation of free eutectic growth at fixed thermal undercooling $\Delta T = T_E - T_O$ in the absence of temperature gradient, where $T_E$ and $T_O$ are respectively the eutectic temperature and the isothermal temperature of the interface. Note that during directional solidification of a single solid phase the presence of a temperature gradient is always necessary to enforce global stability of a cellular array. In its absence one cell would immediately emerge ahead of its neighbors. During eutectic growth however the solid-solid surface tensions between neighboring lamellae are sufficient to maintain the global stability of the array over several lamellae, even in the absence of gradient. It is this property which allows us here to consider the simpler situation of free eutectic growth. In simulations, the direction of growth is then prescribed by two parallel vertical lines with periodic boundary conditions.

The Jackson-Hunt model for free growth predicts a one parameter family of steady states with lamellar spacings and growth velocities related by

$$v = v_m \left\{ \frac{2}{\Lambda} - \frac{1}{\Lambda^2} \right\}, \quad \Lambda = \frac{\lambda}{\lambda_m} \tag{1}$$

where $\lambda_m \sim \Delta T^{-1}$ is the spacing at which the growth velocity reaches its maximum value $v_m \sim \Delta T^2$ (expressions for $\lambda_m$ and $v_m$ are given in Ref. 1). We further restrict our attention to the experimentally relevant limit of small thermal undercoolings where growth occurs slowly and the diffusion equation can be approximated by the Laplacian. The motion of the eutectic interface is then governed by,

$$\nabla^2 u = 0 \tag{2}$$

conservation of mass at both solid-liquid interfaces,

$$V_n = -\hat{n} \cdot \vec{\nabla} u \qquad \alpha\text{–phase} \tag{3a}$$

$$Q\, V_n = \hat{n} \cdot \vec{\nabla} u \qquad \beta\text{–phase} \tag{3b}$$

local thermodynamic equilibrium at the interface (Gibbs-Thomson relations),

$$u(\alpha) = \Delta T\, m_\alpha^{-1} - d_\alpha \kappa \qquad \alpha\text{–phase} \tag{4a}$$

$$u(\beta) = -\Delta T\, m_\beta^{-1} + d_\beta \kappa \qquad \beta\text{–phase} \tag{4b}$$

and the constraint of mechanical equilibrium which requires the sum of surface tensions to vanish at a triple point

$$\sigma_{L\alpha}\, \hat{i}_{L\alpha} + \sigma_{L\beta}\, \hat{i}_{L\beta} + \sigma_{\alpha\beta}\, \hat{i}_{\alpha\beta} = 0 \tag{5}$$

where $\hat{i}_{\gamma\nu}$ is a unit vector which is tangent to the $\gamma - \nu$ interface at a triple point and points away from this point. Here $\sigma_{\alpha\beta}$ is the $\alpha - \beta$ surface tension, $\sigma_{LS}$ the $L - S$ surface tension, $m_S$ is the liquidus slope of each phase $dT/du$ defined to be positive, $d_S = \dfrac{\sigma_{LS}\, T_E}{m_S\, L_S}$ and $L_S$ are respectively the capillary length and latent heat per unit volume of each phase ($S = \alpha, \beta$), $\kappa$ is the local interfacial curvature, $u \equiv (C - C_E)/\Delta C$ is a dimensionless composition field where $C$ denotes the concentration of one of the component, for example the number of B molecules per unit volume for a two-component system consisting of A and B molecules, $C_S$ is the concentration of each solid phase, $\Delta C \equiv C_\beta - C_\alpha > 0$ is the miscibility gap, $C_E$ the eutectic concentration, and $C_\infty$ the concentration in the sample far ahead of the solidification front. In addition we have defined $V_n = (-u_\alpha) v_n / D$ where $v_n$ is the local normal velocity of the interface, $D$ the coefficient of solute diffusivity, and $Q \equiv u_\beta / (-u_\alpha)$ with $u_S \equiv (C_S - C_E)/\Delta C$ ($u_\alpha < 0, u_\beta > 0$). Finally, the exponentially decaying part of $u$ in the upward $z$ direction is translated, in the Laplacian limit of the diffusion equation, into a linear gradient boundary condition on $u$ far ahead of the interface. In steady state:

$$(\partial_z u)_\infty = \frac{v}{D} \, u_\infty$$

$$u_\infty \equiv \frac{C_\infty - C_E}{\Delta C}$$

where mass conservation implies $C_\infty = C_\alpha \eta + C_\beta (1 - \eta)$ and $\eta$ is the volume fraction of the $\alpha$ phase. It is also useful to rewrite the condition of mass conservation in the form $\eta = Q (1 + Q)^{-1} - u_\infty$.

## The Algorithm

We now describe a random walk model which can simulate eqns. 2-5. The model and some results have been briefly described in Ref. 5. Additional results are presented here and a more complete exposition will be given elsewhere (11).

Points on a two-dimensional square lattice are divided into three categories: $\alpha$-sites and $\beta$-sites ($s$-sites; $s = \alpha, \beta$) represent sites occupied by the solid $\alpha$ and solid $\beta$ phases respectively and empty sites ($e$-sites) those occupied by the liquid phase. The lattice spacing is set equal to unity and W measures the lateral width of the system where periodic BC are imposed at the endpoints. The interface, from which walkers are released, is composed of all $s$-sites which have at least one bond connected to an $e$-site (i.e. the solid-liquid interface). The composition field $u$ is interpreted as the probability density of random-walkers which is well known to satisfy Laplace equation. Accordingly, the probability of a walker being released from a particular $s$-site is given by $u(s)$ the value of $u$ on the interface. Since $u(s)$ can be both positive and negative we use the normalized probability distribution $P(s) \equiv |u(s)|/\max\{|u|\}$ and assign to walkers a flux $f(s)$ equal to the sign of $u(s)$ ($f(s) = \pm 1$), where $\max\{|u|\}$ is the maximum value of $|u|$ on the interface. Each $s$-site on the interface is then checked sequentially and a random walker carrying a flux $f(s)$ is released with probability $P(s)$. The walk is terminated when the walker returns to any $s$-site. If a random walker steps above a horizontal line, one lattice site above the highest point on the interface, it is returned along this line according to a probability distribution calculated beforehand (6). This method avoids lengthy walks. The net flux of walkers in all $s-e$ bonds joining $s$-sites and $e$-sites is recorded. For each walk the flux $f(s,e)$ in the $s-e$ bond through which the walker leaves the interface is increased by $f(s)$ and the flux in the $s' - e'$ bond through which it returns is decreased by $f(s)$. Interface motion at site $s$ is controlled by the normal gradient of $u$ at the interface $u(s) - u(e)$ which is proportional to the flux in the corresponding $s-e$ bond ($f(s,e) \sim [u(s) - u(e)]/4$). Accordingly, the sum $\sum_s f(s,e)$ of fluxes in all $s-e$ bonds connected to a given $e$-site is evaluated. This site then becomes occupied by an $\alpha$-site if at least one of the $s-e$ bonds is of type $\alpha-e$ and the sum $\sum_s f(s,e)$ exceeds an integer $M$, or by a $\beta$-site if at least one of the bonds is of type $\beta - e$ and $\sum_s f(s,e)$ is less than $-QM$ (see eqns. 3a-b). Similarly, an $s$-site is emptied (melting) when the sum $\sum_e f(s,e)$ of fluxes in all bonds connected to this site is less than $-M$ for $s = \alpha$ or exceeds $QM$ for $s = \beta$. When $M$ is large many walks take place during the time necessary for the interface to move one lattice unit and fluctuations in the normal gradient of $u$ at the interface are therefore diminished. In the limit $M \to \infty$ deterministic equations of motion are simulated.

The effect of the boundary-layer of excess concentration at off-eutectic melt composition is simulated by releasing random walkers from infinity. More specifically $M_\infty$ random walkers, each carrying a flux equal to the sign of $u_\infty$, are released from a horizontal line above the interface each time one new lattice site becomes occupied (i.e., an $e$-site becomes an $s$-site). In the converse process where an occupied site is emptied the sign of the flux carried by these walkers is reversed. In the algorithm the off-eutectic melt composition is then controlled by $M_\infty$ where:

$$|u_\infty| = \frac{M_\infty}{M} \cdot \frac{1}{(1+Q)}$$

Recall that $Q \equiv C_\beta - C_E/(C_E - C_\alpha)$. It should be noted that this way of incorporating in the algorithm the effect of the boundary-layer can only describe situations where the average vertical velocity of the eutectic front remains constant (11). This does not require, however, the front to grow in steady-state. The solid-liquid interface shape can vary in time, the average vertical velocity remaining constant. This limitation is not present in simulations performed exactly at eutectic composition ($u_\infty = 0$).

Three types of walks can be distinguished and each type given clear physical interpretation. Walks that start from an $s$-site and end on an $s$-site of the same phase ($\alpha - \alpha$ and $\beta - \beta$ walks) do not contribute to net interface motion, since they conserve flux and only displace material. They simulate the effect of capillary forces at the solid-liquid interface. On the contrary $\alpha - \beta$ and $\beta - \alpha$ walks correspond to diffusion between neighboring lamellae of rejected A and B molecules and contribute to the net growth of both solid phases. Finally walkers released from infinity at off-eutectic compositions enhance the growth of one solid phase preferentially.

The last task is to build in the model the constraint of mechanical equilibrium (eqn. 5). To do so we derive a form of the Gibbs-Thomson relation which takes into account the interaction energy between the three phases at triple points. Away from these points this form reduces to eqns. 4a-b. In their vicinity, that is within a distance $r$ of triple points where $r$ measures the range of the interaction energy between different phases, this form causes the composition field $u(s)$ to become large when slight deviations from the constraint of mechanical equilibrium occur. This increase in composition in turn induces large normal composition gradients which cause interface motion to smooth out these deviations in a time much shorter than the time it takes for triple points to move a distance $r$. Consider the thermodynamic potential corresponding to the interfacial region $\Omega_r = \sum_s \Omega_r(s)$ where:

$$\Omega_r(s) = -\mu_\alpha N_\alpha - \mu_\beta N_\beta - \mu_L N_L + a_{\alpha\beta} N_\alpha N_\beta + a_{\alpha L} N_\alpha N_L + a_{\beta L} N_\beta N_L,\tag{6}$$

$\mu_\nu$ is the chemical potential of each phase ($\nu = \alpha, \beta, L$), $N_\nu$ the number of lattice sites of each phase contained inside a circle of radius $r$ centered on site $s$, and $a_{\gamma\nu}$ measures the strength of the interaction energy between the $\gamma$ and $\nu$ phases. The corresponding Gibbs-Thomson relation is then obtained by imposing $\Omega_r$ to be stationary against infinitesimal interface deformations equivalent to adding or withdrawing a surface site: $\partial\Omega_r(s)/\partial N_\alpha = \partial\Omega_r(s)/\partial N_\beta = 0$, with the constraint $N_\alpha + N_\beta + N_L = N_r$ where $N_r$ is the total number of sites inside the circle of radius $r$ ($N_r = \pi r^2$ for $r \gg 1$), and further using the fact that $u(\alpha) - m_\alpha^{-1} \Delta T \propto \mu_L - \mu_\alpha$ and $u(\beta) + m_\beta^{-1} \Delta T \propto \mu_\beta - \mu_L$. We obtain

$$u(\alpha) - m_\alpha^{-1} \Delta T = -a_{\alpha L} (N_L - N_\alpha) + (a_{\beta L} - a_{\alpha\beta}) N_\beta\tag{7a}$$

$$u(\beta) + m_\beta^{-1} \Delta T = a_{\beta L} (N_L - N_\beta) - (a_{\alpha L} - a_{\alpha\beta}) N_\alpha.\tag{7b}$$

where the constants of proportionality between $\mu_L - \mu_S$ and $u(s) \pm m_S^{-1} \Delta T$ have been absorbed by suitably redefining the $a_{\gamma\nu}$'s and eqns. 7a-b determine $P(s)$. Away from triple points the second terms on the right-hand sides of eqns. 7a-b vanish and $N_L - N_S$ is proportional to the interfacial curvature: $\kappa = B(r) (N_L - N_S)/r^3$ where $B(r) \to 3/2$ for $1 \ll r \ll \kappa^{-1}$. Eqns. 7a-b therefore reduce to the form of eqns. 4a-b. Finally, for $u(s)$ to be continuous at a triple point the right-hand sides of eqns. 7a-b

40

have to scale as $r/\lambda$ and vanish in the limit $r/\lambda \to 0$. This requirement and the constraint $N_\alpha + N_\beta + N_L = N_r$ then fix uniquely $N_\alpha, N_\beta$, and $N_L$ within a circle of radius $r$ centered at this point, and consequently the directions of the $\hat{t}_{\gamma v}$'s, in terms of the $a_{\gamma v}$'s (which also relates the $a_{\gamma v}$'s with the $\sigma_{\gamma v}$'s). Modified versions of eqns. 7a-b are used in simulations where $\lambda$ is only a few times $r$:

$$u(\alpha) = m_\alpha^{-1} \, \Delta T - \frac{a\,B(r)}{r^3} \, \{(N_L - N_\alpha)\,(1 + 2r^{-1}N_\beta) + (2r + 1)\} \qquad (8a)$$

$$u(\beta) = -m_\beta^{-1} \, \Delta T + \frac{a\,B(r)}{r^3} \, \{(N_L - N_\beta)\,(1 + 2r^{-1}N_\alpha) + (2r + 1)\} \qquad (8b)$$

where we have restricted our attention to the special case of equal surface tensions and defined: $a\,B(r)/r^3 \equiv a_{\alpha\beta} = a_{\alpha L} = a_{\beta L}$. An additive factor of $2r + 1$ has been introduced to insure that the curvature of a flat interface away from triple point vanishes (i.e. $N_L - N_\alpha = -(2r + 1)$ for a flat interface since the circle of radius $r$ is centered on an $S$-site), and a multiplicative factor $1 + 2r^{-1}N_s$ to reinforce the constraint of mechanical equilibrium at the triple points. For this constraint to remain satisfied during interface motion the jump in composition $\delta u_0$ due to a slight shape perturbation in the vicinity of a triple point has to be much larger than the size of typical composition variations $\delta u_i$ along the interface. Since in the absence of a multiplicative factor, $\delta u_o$ scales as $1/r$ while $\delta u_i$ scales as $1/\lambda$, this condition is only satisfied when $\lambda >> r$. When $\lambda$ is only a few times $r$, which is the case in simulations with small lattices or several lamellae, the contact angles at triple points fluctuate significantly. The multiplicative factor $1 + 2r^{-1}N_s$ is equal to unity away from triple points, which insures that eqns. 8a-b reduce to eqns. 4a-b, and makes $\delta u_0$ of order unity which insures $\delta u_0 >> \delta u_i$ and these angles to remain fixed during growth.

It should be emphasized that this random-walk model simulates the deterministic continuum eqns. 2-5 only when both limits $M >> 1$ (deterministic) and $1 << r << \lambda$ (continuum) one satisfied simultaneously. In simulations values of $M$ in a range between 10 and 20 are typically sufficient for the interface evolution to be independent of the sequence of random numbers used to generate the walks, apart from noise induced global symmetry breaking (i.e. left-right symmetry of the "tilting mode", displayed in Fig. 3a), and therefore deterministic for all practical purposes. For steady-state growth, we have compared the results of simulations, performed on large lattices ($W = 240$, $r = 10$, $M = 15$) using eqns. 7a-b, with the prediction of the Jackson-Hunt analysis. A good quantitative agreement between the two was found, the discrepancy (10 - 20%) being of the same magnitude as the error introduced by the approximations used in the Jackson-Hunt analysis. A more accurate quantitative test of the continuum limit of the algorithm would require an exact solution to the steady-state equations and using even larger lattices. Finally, because of limited computation time simulations with several lamellae are performed with eqns. 8a-b and values of $r$ and $\lambda$ ($r \simeq 5, \lambda \simeq 20 - 60$) which introduce larger corrections to the continuum limit. Our hope is that these corrections only have quantitative effects and do not change the qualitative dynamical behavior of the model. This was found to be true in all cases where a simulation was repeated with larger values of $\lambda$ and $r$.

## Morphological Instabilities

In this study we have restricted our attention to an idealized system with a symmetrical binary eutectic phase diagram with $m_\alpha = m_\beta \equiv m$, $Q = 1$ and equal surface tensions $a_{\alpha L} = a_{\beta L} = a_{\alpha\beta}$. All our simulation results are parametrized in terms of $u_\infty = (C_\infty - C_E)/\Delta C$ and $\Lambda = \lambda/\lambda_m$. When the lamellar spacing is too small ($\Lambda < 1$) an instability leading to termination of lamellae develops. A simulation exhibiting this termination process is shown in Fig. 2. This instability has been widely observed experimentally (1,12) and modeled analytically in terms of a long-wavelength phase equation introduced by Langer (13). On the other hand, if the lamellar spacing is too large a liquid pocket develops within individual lamellae. This phenomenon was observed in our simulations and has also been commonly observed experimentally.

41

As a consequence of these two processes occurring when $\Lambda$ is either too small or too large steady-state growth is limited to a range of lamellar spacings. At off-eutectic melt compositions steady-state growth is even further restricted by the appearance of additional instabilities: the oscillatory instability at twice $\lambda$ predicted successfully by the discrete stability analysis of Dayte and Langer (Fig. 3b), and a new tilting instability on the lamellar spacing which forces lamellae from the phase of smaller volume fraction ($\beta$-phase here) to bend coherently on one side of the vertical growth axis (Fig. 3a).

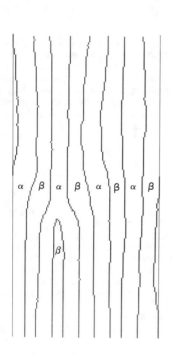

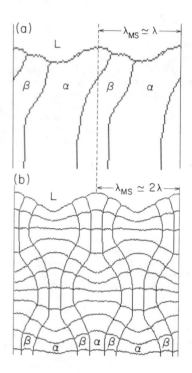

Fig. 2    Simulation exhibiting the termination of one lamella.

Fig. 3    Interfacial deformations associated with (a) tilting instability and (b) oscillatory instability. Lamellae bend in response to a bulge in the interface to preserve balance of surface tensions at triple points. In (a) $W = \lambda = 61$ and $\Lambda = 1.8$, (b) $W = 2\lambda = 121$ and $\Lambda = 1.1$, and in both (a) and (b) $M = 15$, $r = 5$, and $u_\infty = -1/6$. (b) shows only the late stages of a simulation that started from slightly perturbed steady-state growth. Tilting can either be right or left of the vertical growth axis.

At fixed $\Lambda$ the tilting instability first occurs at a more off-eutectic composition (larger value of $|u_\infty|$) than the oscillatory one, and at fixed composition it occurs at a larger value of $\Lambda$. In simulations with $u_\infty = -1/6$ and $W = 2\lambda$ oscillations terminate steady-state lamellar growth when $\Lambda > \Lambda_{OS} \simeq 1$ while in simulations with the same value of $u_\infty$ and $W = \lambda$ steady-states become unstable against tilting when $\Lambda > \Lambda_{TI} \simeq 1.5$.

The appearance of both oscillatory and bending modes can be explained physically by considering the destabilizing effect of the composition gradient ahead of the interface $(\partial_z u)_\infty = v u_\infty / D$, present at off-eutectic compositions. First, note that in the presence of such a gradient an initially planar single solid phase ($\alpha$ phase here if we treat the lamellar eutectic as a single phase) will develop during its growth a spatially sinusoidal deformation of wavelength $\lambda_{MS}$ as a result of the classic Mullins-Sekerka instability (14). Then, equating the spatial period of the interface deformation corresponding to each mode with $\lambda_{MS}$ (as shown in Figs. 3a-b) yields the criterion that oscillations and tilting should first occur when $\lambda_{MS} = 2\lambda$ and $\lambda_{MS} = \lambda$ respectively. Since $\lambda_{MS}$ is proportional to $(d_\alpha D / v |u_\infty|)^{1/2}$ and decreases as $|u_\infty|$ increases (i.e., a steeper gradient or thinner boundary layer induces an instability on a shorter length scale) the tilting instability should occur, according to this criterion, at a more off-eutectic melt composition than the oscillatory instability (i.e., the value of $|u_\infty|$ required for $\lambda_{MS} = \lambda$ to be satisfied has to be larger than the one required for $\lambda_{MS} = 2\lambda$ to be satisfied). This is at least in qualitative agreement with our simulations. Reliable quantitative predictions however can not be expected from this crude criterion.

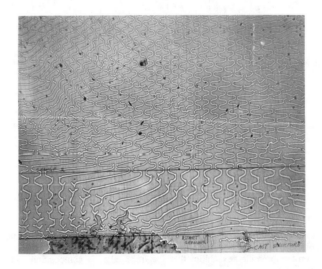

Fig. 4     Experimental photograph by Kaukler of an oscillatory eutectic morphology in carbon-tetrabromide-hexachloroethane. The horizontal width corresponds approx. to 1.25 mm.

We find in our simulations that a rich dynamical behavior can take place above onset of short wavelength instabilities, within some range of growth rates and compositions. In particular, we observe that when either instability (oscillatory or tilting) dominates the dynamics coherent structures are formed (5), while in regions where both instabilities compete the motion of lamellae can become chaotic. We shall not attempt here to describe this behavior in great detail, but present experimental evidence for the existence of non-typical eutectic morphologies, similar to the ones obtained in simulation.

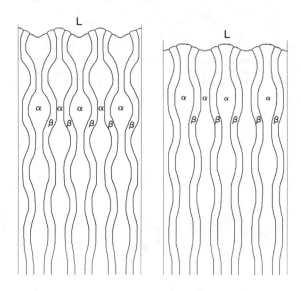

Fig. 5         Simulated oscillatory
                eutectic morphologies.

Fig. 4 shows an experimental photograph by Kaukler of an oscillatory eutectic morphology (3). This morphology was formed by directionally solidifying a very thin film (thickness is about 10-20 μm) of carbon-tetrabromide-hexachloroethane at near eutectic composition (8.635 w/o of hexachloroethane).

For comparison we have displayed in Fig. 5 two simulated eutectic morphologies generated above threshold of the oscillatory instability ($u_\infty = -1/6$). The one with larger amplitude oscillations correspond to a slightly faster growth rate.

Although the binary phase diagram and material parameters corresponding to the idealized system studied by simulation differ from the one of carbon-tetrabromide-hexachloroethane, both the simulated and experimentally observed morphologies share the same oscillatory character. Other features of the experimentally observed morphology, such as the tendency for lamellae of the phase of smaller volume fraction to branch and follow oblique path (see bottom of Fig. 4) are likely to be consequences of the tilting instability (11).

44

There seems to be sufficient motivation at this point to perform additional systematic experimental studies of off-eutectic solidification of very thin films, where three dimensional effects are essentially suppressed. These experiments would explore further the range of possible microstructures, and, more fundamentally, investigate complex spatiotemporal behaviors ranging from orderly to chaotic.

## Future Prospects

We have only considered so far eutectic solidification of a thin film in the limit of slow velocity. Of course, much more work is now needed to fully understand this system but the remarkable agreement between some of the simulated and experimentally observed eutectic morphologies is at least encouraging. Probably the most promising outlook for the near future is to extend lattice simulations to three dimensions where questions of pattern selection involving the motion of lamellar faults remain poorly understood. It is also conceivable that the model discussed in this paper could be modified to investigate the even vaster class of eutectic microstructures which form when the solid-liquid interface of one phase is faceted (15).

## Acknowledgements

I wish to thank M.C. Cross for many helpful discussions. I am also grateful to Prof. J.S. Langer for communicating to me the experimental works of J. Van Suchtelen and W.F. Kaukler, and to W.F. Kaukler for providing additional helpful information and for allowing me to include one of his unpublished experimental photographs in this manuscript. This research was supported by the California Institute of Technology through a Weingart Fellowship and through the Program in Advanced Technologies which is funded by GM, GTE, TRW and Aerojet.

## References

1.    K.A. Jackson and J.D. Hunt, Trans. Metall. Soc. AIME $\underline{236}$ , 1129 (1966).

2.    D.P. Woodruff, The Solid-Liquid Interface (Cambridge University Press, 1973).

3.    W.F. Kaukler (private communication).

4.    J. Van Suchtelen (unpublished).

5.    A. Karma, Phys. Rev. Lett. $\underline{59}$ , 71 (1987).

6.    L.P. Kadanoff, J. Stat. Phys. $\underline{39}$ , 267 (1985).

7.    C. Tang, Phys. Rev. $\underline{A31}$ , 1977 (1985).

8.    S. Liang, Phys. Rev. $\underline{A33}$ , 2663 (1986).

9.    D.T.J. Hurle and E. Jakeman, J. Cryst. Growth $\underline{3, 4}$, 574 (1968).

10.    V. Dayte and J.S. Langer, Phys. Rev. $\underline{B24}$ , 4155 (1981).

11.    A. Karma, (in preparation).

45

12.  V. Seetharaman and R. Trivedi (to be published).

13.  J.S. Langer, Phys. Rev. Lett. $\underline{44}$ , 1023 (1980).

14.  W.W. Mullins and R.F. Sekerka, J. Appl. Phys. $\underline{35}$ , 444 (1964).

15.  D.J. Fisher and W. Kurz, *Proc. Quality Control of Engineering Alloys and the Role of Metals Science* (edited by H. Nieswaag and J.W. Schut) p. 59, Dept. of Met. Sci. & Tech., University of Delft (1977).

# INFLUENCE OF CONVECTION ON EUTECTIC MORPHOLOGY

P.A. Curreri, D.J. Larson*, and D.M. Stefanescu**

NASA, Marshall Space Flight Center, AL
*Grumman Corp. Research Center, Bethpage, NY
**The University of Alabama, Tuscaloosa, AL

ABSTRACT

Experimental data is given for BiMn/Bi rod eutectic and
$Fe_3C$/Fe lamellar eutectic solidified in low gravity. Eutectic
spacing is summarized for various alloy systems solidified in
low-gravity. On-eutectic and off-eutectic models are evaluated
with respect to the low-gravity solidification data. The models
examined are inadequate for quantitative prediction of phase
spacing for on-eutectic solidification in low gravity.

Solidification Processing of Eutectic Alloys
D.M. Stefanescu, G.J. Abbaschian and R.J. Bayuzick
The Metallurgical Society, 1988

## Introduction

Solidification processes are strongly influenced by gravitational acceleration (1). Consider a simple plane front single-phase binary alloy. Convective flow mixes solute in the liquid ahead of the solid-liquid interface preventing diffusion limited steady state growth. The result is continually varying solid composition and properties. Gravity has additional influence in dendritic growth. Dendrite morphology, in transparent organic models, depends on directional orientation to gravity (2). A complex balance between intradendritic thermal and solutal density gradients controls convection up flows that cause "freckle" type casting defects (3,4). Stokes flow of second phase particles in off-eutectic and off-monotectic alloys, and noncontiguous metal matrix composites severely limits casting composition.

Buoyancy independent solidification within earths gravitational field is accomplished only within strict limits. In one dimension, strong magnetic fields can dampen convection (5), and density gradients can be oriented with gravity for stability. But, magnetic flow damp in one direction increases flow velocity (segregation, etc.) in the transverse direction (6). Opposition of thermal and solutal convection for many alloy compositions make stabilization of convection by orientation, even in one dimension, unfeasible.

Space flight provides solidification research the first long duration access to micro-gravity (7). Supporting commercial and academic interest in solidification in Space are several short duration free fall facilities. These include: drop towers (4 s, 0.0001 g, g=980 cm/s$^2$), parabolic aircraft flight (30 s, 0.01 g), and suborbital sounding rockets 5 min., 0.0001 g).

Melt segregation and the perturbation of diffusion processes by thermal and solutal convection is negligible in low-gravity Space (8). Metal alloy dendrite primary (9) and secondary (10,11,12) spacing is larger for equivalent solidification conditions except low gravity. Microstructure and phase composition resulting from low-gravity solidification of insitu composite alloys can yield magnetic (13,14) and electrical (15,16) property enhancement. A most unexpected finding (12,13,14,17,18) is that directional solidification of on-eutectic alloy in low gravity often dramatically changes eutectic interphase spacing, as well as eutectic grain size in equiaxed eutectic growth.

This paper examines current understanding of the influence of convection on planar eutectic growth. Low-gravity eutectic solidification results are presented for MnBi/Bi and Fe-Fe$_3$C on-eutectic alloys. Proposed theoretical models are discussed, in relation to low-gravity solidification, for eutectic phase spacing.

## Experimental Methods

Directional solidification for MnBi/Bi and Fe/Fe$_3$C eutectics was accomplished in similar Bridgman-Stockbarger type

48

furnaces. Two versions of the automatic directional solidification system (ADSS) built by G.E. for NASA were used - Space Application Rocket Program version (for MnBi-Bi) and ADSS prototype modified for KC-135 parabolic aircraft flight (for $Fe/Fe_3C$). The liquid thermal gradients at the solid-liquid interface, $G_L$, were 100 $\pm$3 C/cm (MnBi/Mn) and 201 $\pm$20 C/cm ($Fe/Fe_3C$).

On-eutectic alloys were prepared by induction melting. MnBi/Bi eutectic samples were prepared from commercially pure Mn (99.9 w/o) and high purity Bi (99.999 w/o). Bulk composition and uniformity were determined using differential scanning calaorimetry and chemical sperophotometric absorbance. The eutectic composition (19) was 0.72 $\pm$0.09%. $Fe/Fe_3C$ samples were prepared from laboratory purity composition. Typical chemical analysis (wt. %) was 4.,28 C, 0.01 Si, 0.016 Mn, 0.003 P, 0.003 S, yielding a carbon equivalent of 4.28.

SPAR (MnBi/Bi) sounding rocket samples, 5 cm. long 0.4 cm. di., were directionally solidified in quartz ampoules for about 4.5 minutes of low gravity (0.0001 g). Samples were soaked at temperature for 90 minutes preflight. Directional solidification commenced 15 s after 0.0001 g was obtained. Eighty percent of the sample was solidified in the low-gravity portion of the flight and the remainder during rocket re-entry (20). KC-135 ($Fe/Fe_3C$) samples, 7 cm. long, 0.5 cm. di. were directionally solidified in alumina crucibles during multiple low gravity maneuvers. Prior to parabolic flight the samples are thermally soaked for 10 minutes at furnace temperature of 1500 C. About two cm. of sample is left unmelted allowing reliable identification of the parent melt interface. Each maneuver consisted of 20-30 seconds of low-gravity (0.01 g) and 1 to 1.5 minutes of high- gravity (1.7 g) forces parallel to the longitudinal growth axis. For a typical maneuver during low-g the acceleration on all axes averages below 0.01 g. During pullout and climb, the high-g acceleration parallel to the longitudinal axis reaches 1.75 while the accelerations on the other axes are less than 0.15 g. The known sample solidification rate is then correlated with accelerometer data to determine the gravity level during solidification for any position on the sample (12).

## Results

### MnBi/Bi Fibrous Eutectic:

On-eutectic MnBi/Bi was solidified in low gravity on SPAR VI (R=6 mm/min) and SPAR IX (8.3 mm/min) (20). Sample microstructure, composition, and properties were compared relative to 1 g controls. The 1 g controls were solidified under identical conditions except for gravity. The eutectic interphase spacing (relative to 1 g controls) for low-gravity solidified sample decreases by over 50%. This is evident in the transverse sections given in Fig. 1. A decrease in rod diameter is also apparent in Fig. 1 for low-gravity solidified sample. Volume fraction of MnBi rods in low gravity samples is also smaller (about 7%). Thermal data reveals increased interfacial undercooling (3-5 C) during low-gravity solidification. Table I summarizes these results. Low gravity

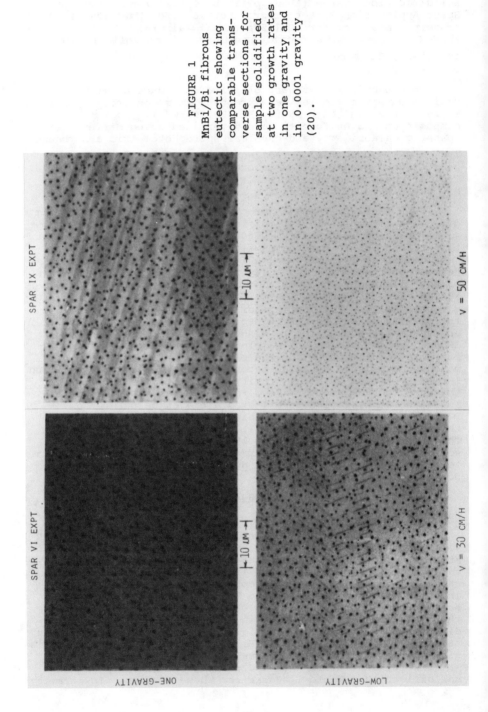

FIGURE 1
MnBi/Bi fibrous
eutectic showing
comparable trans-
verse sections for
sample solidified
at two growth rates
in one gravity and
in 0.0001 gravity
(20).

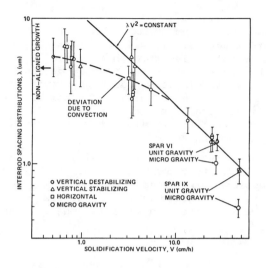

FIGURE 2
Interrod spacing versus
solidification velocity
for MnBi/Bi grown under
low gravity and under
various orientations in
one gravity (13).

solidification produced samples with increased intrinsic
coercivity (resistance to demagnetization). Physical
properties will not be discussed in this paper.

Table I:  Thermal and Morphological Properties of MnBi/Bi

| Gravity level<br>Growth Rate | low gravity<br>6 | <br>8.3 | 1 gravity controls<br>6 | <br>8.3 mm/min |
|---|---|---|---|---|
| Eutectic<br>Spacing (microns) | 1.00 | 0.48 | 1.54 | 0.90 |
| Rod Diameter<br>(microns) | 0.33 | 0.18 | 0.53 | 0.33 |
| Volume Fraction<br>Change (%MnBi) | $-7^{\pm}3$ | $-8^{\pm}3$ | ---- | ---- |
| Solid/liquid<br>Interface<br>Undercooling (C) | $2.75^{\pm}2$ | $5.5^{\pm}2.7$ | ---- | ----- |

Phase spacing refinement can of course also be achieved in
one gravity by increasing solidification rate, R.  Fig. 2 gives
the interrod spacing as a function of R under various processing
conditions.  It is evident (Fig. 2) that orientation of sample
in one gravity has no effect on λ.  At higher R the spacing
appears to obey $\lambda R^2$ = constant (21), with the exception of
finer spacing for low-gravity samples.

Fe/Fe₃C Lamellar Eutectic:

    On-eutectic Fe/Fe₃C (austenite-cementite) directionally
solidified with a lamellar structure. Interlamellar spacing is
given in Fig. 3 for a sample solidified through multiple
low-gravity maneuvers. The spacing decreases sharply during
solidification in low gravity, and increases in high-gravity.
A very pronounced coarsening of the structure occurs toward the
middle of some high-gravity sections, as shown in Fig. 4 for
two sections. Refinement of lamellar spacing after 25 seconds
low gravity is typically 25 percent.

    Interphase spacing refinement during low-gravity
solidification is observed for both fibrous (MnBi/Bi) and
lamellar (Fe₃C/Fe) on-eutectic compositions. Table II
includes these results and data reported by other authors.
Although results for each alloy prove reproducible, there is no
obvious trend.

Table II

Effect of Low Gravity on On-Eutectic Interphase Spacing
Various Authors)

| Alloy Composition | Low-Gravity Solidification Effect on Interphase Spacing |
|---|---|
| Lamellar Eutectics | |
| Al₂Cu/Al (22,17) | no change |
| Fe₃C/Fe | − 25 % |
| Fibrous Eutectics | |
| MnBi/Bi | − 50 % |
| InSb/NiSb (18) | − 20 % |
| Al₃Ni/Al (17) | + 17 % |

Discussion

    The mechanism for the influence of low gravity during
solidification on eutectic phase spacing is not readily
apparent. The importance of convective flow is demonstrated
for MnBi/Bi (4) , by the duplication of low gravity spacing and
interfacial undercooling through solidification in strong (3
kG) magnetic fields. The effect of gravity on convective flow
is relatively well understood. The challenge is to develop a
theory adequately relating convection flow at low gravity to
the mechanisms controlling eutectic phase spacing. Several
approaches recently suggested to develop the needed theory are
examined in relation to experimental findings.

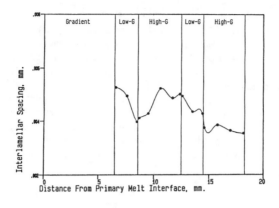

FIGURE 3

Variation of the interlamellar spacing of Fe$_3$C/Fe
eutectic directionally solidified through multiple
low-g (high-g) parabolas. Solidification rate = 5.17
mm/min (12).

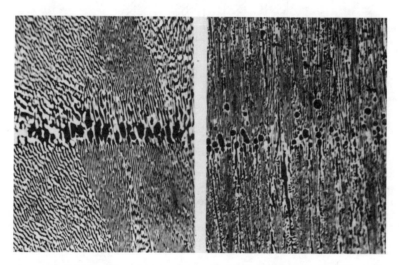

a) First high-g zone          b) Fourth high-g zone

SAMPLE 84-F-3 (Pure Fe-C alloy), R=5.17 mm/min.
Bands in high-g zones, 200x
Solidification direction upward

FIGURE 4

Fe$_3$C/Fe lamellar eutectic showing bands of coarser
eutectic in the middle of some high-g zones. Magnification 200 times (12).

53

# Analysis of Convective Flow for On-Eutectic Growth

Simple stagnant film model predicts (1,23) that cooperative eutectic growth will be unaffected by the presence (or absence) of convection. The model considers growth taking place in a boundary layer in which solute transport is by diffusion only. Outside this boundary layer the liquid is mixed by convective flow. Solute redistribution for on-eutectic growth occurs on the scale of the lamellar spacing, which is much smaller than the convection flow momentum boundary layer. Thus, for steady state eutectic growth at fixed volume fraction, convection flow is not expected to affect lamellar spacing.

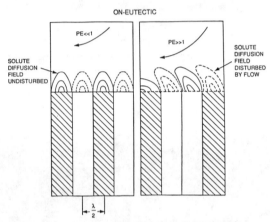

FIGURE 5

Schematic isoconcentration curves ahead of the growth interface for on-eutectic showing diffusion fields disturbed for high Pe and undisturbed for small Pe (24).

Quenisset and Naslain (24) apply a more rigorous analysis and demonstrate invalidity in the above conclusions. The relative effect of convection to the diffusion field for on-eutectic growth is given by the Peclet number, Pe, with the characteristic dimension of the phase spacing.

$$Pe = U \, \lambda \, / \, 2 \, D_L \tag{1}$$

U is the characteristic velocity of the flow across the solidification interface, and $D_L$ is the liquid diffusion coefficient. If Pe is small, the concentration field in the liquid is not disturbed by convection. If Pe is large, convection disturbs the concentration fields and affects cooperative eutectic growth (Fig. 5). The actual flow fields (diffusive and convective) in the liquid ahead of the eutectic solid-liquid interface are numerically calculated. A series of curves for lamellar spacing versus interfacial undercooling for a given R (Fig. 6) can be determined for different magnitudes of forced convective flow.

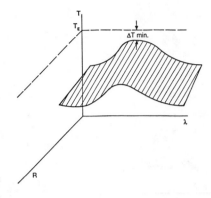

FIGURE 6
Schematic of solidification at the "extremum" - at the interphase spacing giving the minimum solid-liquid interface undercooling for a given solidification rate (21).

Like the approach of Jackson and Hunt (21) Zener's assumption that growth is preferred at the "extremum" (minimum interfacial undercooling, $\Delta T$) defines the eutectic spacing for a given R. The interfacial undercooling is assumed to be small so that the average solid composition, $C_{(x)}$, equals the eutectic composition, $C_E$.

The predictions of the theory are summarized in Fig. 7. Forced convection decreases interfacial undercooling and increases interphase spacing. The theory semiquantitatively predicts phase spacing at high convection flows for $Ti/Ti_5Si_3$ and $Fe/Fe_2Bi$. Qualitatively, the theory predicts the low-gravity results for MnBi/Bi eutectic (Table I).

Baskaran and Wilcox (25) apply the Quenisset and Naslain analysis to eutectic MnBi/Bi. Crucible rotation is used to provide forced convective flow. The experimental MnBi/Bi spacing increase due to forced convective flow is semiquantitatively predicted by the theory.

The theory is then applied to MnBi/Bi low-gravity solidification data (Table I). The convective flow in the Bridgman-Stockbarger flight furnace in one gravity and 0.0001 gravity is calculated (26). The calculated disturbance in the eutectic diffusion field at one gravity is so slight that the calculated eutectic phase spacing for solidification in one gravity or in Space are essentially equivalent. The inadequancy of the theory to predict eutectic spacing data for low gravity is shown schematically in Fig. 8. Thus, the calculated effect of convective flow on the liquid concentration field does not explain the low-gravity eutectic phase spacings.

Off-Eutectic Approach

Favier and deGoer (17) propose an off-eutectic model to account for the influence of low gravity on eutectic spacing. They postulate an arbitrary 1% deviation from eutectic composition. Boundary layer theory is used to predict the effect of convection on the volume fraction term in the equations for eutectic spacing (21,23).

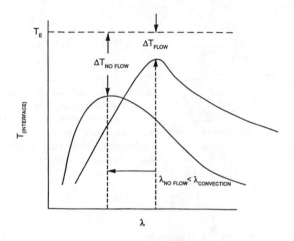

FIGURE 7

Predicted evolution of undercooling-interphase spacing
with forced convective flow parallel to the solid-
liquid interface from the Quenissett-Naslain model (24).

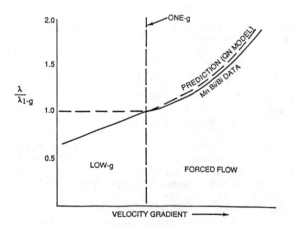

FIGURE 8

Schematic showing the discrepancy between
the Quenisset-Naslain model prediction and
experimental data for MnBi/Bi (26).

Off-eutectic models offer several advantages. The solute redistribution boundary layer is on the order of D/R - much larger than that ($\lambda$) for on-eutectic cooperative growth. Thus, convection has a much more pronounced effect on the concentration field ahead of the solid-liquid interface. The sign of the composition deviation from eutectic determines the sign of the low-gravity induced phase spacing change, i.e., low-gravity spacing that are smaller for hypereutectic and larger for hypoeutectic. No change, insensitivity to convection, is also predicted for some materials.

Essentially, Favier and deGoer's model employs the Jackson and Hunt expression for off-eutectic (plain front) lamellar spacing written in terms of eutectic phase volume fraction, $f_v$ (21,23). Assuming that the difference in $\lambda$ for solidification in one gravity, $\lambda_{1g}$, from that for low gravity, $\lambda_{0g}$ is due only to the effect of convection on $f_v$, an equation of the following form can be written.

$$\lambda_{1g}/\lambda_{0g} = \left[ (f_{v1g} \ A + 1)/(f_{v0g} \ A + 1) \right]^{\frac{1}{2}} \qquad (2)$$

Where $A = (a_\alpha^L/a_\beta^L) \ (m_\beta^L/m_\alpha^L) - 1$; $m_i$ is the liquidus slope of phase i; $a_i$ is a function of melting temperature, surface free energy, heat of fusion, and interphase angle for the eutectic phases as defined in Ref. 21. Favier and deGoer relate $f_v$ to the average solid composition, $C_s$, by

$$f_v = (C_s - C_\alpha)/(C_\beta - C_\alpha) \qquad (3)$$

where $C_\alpha$ and $C_\beta$ are the limit of solubility at the eutectic temperature. $C_s$ is expressed with and without convection using stagnant film theory and substituted into combined Eqs. 2 and 3.

The model predicts 30-50% difference between $\lambda_{1g}$ and $\lambda_{0g}$, if $A = 100$, which may be appropriate for phase diagrams with large dissymmetry (i.e., BiMn/Bi, $Al_3Ni/Al$, InSb/NiSb). Low-gravity spacing is larger than that in one gravity for 1 wt.% hypoeutectic; 1 gravity spacing is larger for 1 wt.% hypereutectic. If $A = 0$, which may be appropriate for symmetrical phase diagram systems as $Al_2Cu/Cu$, then the spacing is not sensitive to convection.

Several objections are apparent concerning the above model. 1. No mechanism is given explaining the composition shift to off-eutectic. Random experimental error in initial composition is not plausible, since all low-gravity $\lambda$ values (Table II) for each alloy composition are repeatable. 2. Apparently equations valid only for lamellar eutectics (21) have been used (17) to compare convection effects of lamellar and rod eutectics. 3. It is not clear that Eq. 3 is valid, since it ignores the phase specific volumes.

In the next section the off-eutectic model is tested against low-gravity results for MnBi/Bi and $Fe_3C/Fe$. A mechanism is suggested for alloy dependent composition shift in low gravity. An expression is given equivalent to Eq. 2 but valid for rod eutectics. The need for use of Eq. 3 (and of

57

stagnant layer theory) is eliminated through use of flight $f_v$ data.

## MnBi/Bi: Rod Eutectic

a) Off-eutectic like solidification due to undercooling in low-g

MnBi/Bi eutectic samples experience considerable solid-liquid interfacial undercooling during directional solidification in low gravity (Table I). Under equivalent solidification conditions except in one gravity MnBi/Mn essentially solidifies at the temperature, $T_E$ (Fig. 9). Undercooling at the eutectic solid-liquid interface is expressed following Jackson and Hunt (21) as

$$\Delta T = \Delta T_C + \Delta T_s + \Delta T_k \qquad (4)$$

$\Delta T_k$, the kinetic term, can usually be ignored for metallic eutectics. $\Delta T_s$, the surface term, is not likely a function of gravity. The gravity dependent term is then $\Delta T_C$, the composition term. $\Delta T_C$ is expressed as:

$$\Delta T_C = m[C_E - C(x)] \qquad (5)$$

where 1/m is the reciprocal sum of the liquidus slopes. Extension and bisection of the liquidus slopes for interfacial undercooling of MnBi/Bi (Fig. 9) give nonequilibrium eutectic composition, $C_E'$, at a lower Bi concentration than $C_E$. Thus, for liquid composition $C_L = C_E$, MnBi/Bi in low gravity solidifies relative to the undercooled phase diagram as a Bi rich off-eutectic (Fig. 10). The average solid concentration is now $C_E$ times a partition coefficient. Volume fraction MnBi data (Fig. 11) for MnBi/Bi show eutectic type solidification in normal gravity and Bi rich off-eutectic type solidification in low-gravity.

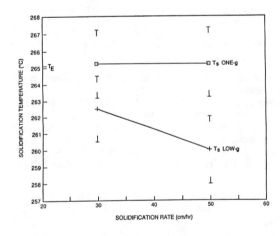

FIGURE 9

Solidification temperature (undercooling) data for MnBi/Bi solidified in low gravity relative to one-gravity controls (20).

It is not known, due to lack of published data, if the other alloys in Table II also experience increased eutectic solid-liquid interface undercooling in low gravity. Alloy dependent undercooling in low gravity could explain the data in Table II. To test this hypothesis, the undercooling in low gravity and the relevant low gravity phase diagrams need to be studied.

b) Test of off-eutectic model with volume fraction flight data

The off-eutectic model can be tested with the $f_v$ flight data for MnBi/Bi (Fig. 11). It should be noted that the mechanism responsible for the difference between $f_{v0g}$ and $f_{v1g}$ in MnBi/Bi (Fig. 11) differs from that proposed by Ref. 17. Ref. 17 assumes off-eutectic solidification with convection (1g) and without convection (low-g). Fig. 11 indicates on-eutectic solidification with convection (1-g) and off-eutectic solidification without convection (low-g). The use of the boundary layer model in Ref. 17 to calculate $f_{v0g}$ and $f_{v1g}$ is replaced with the actual $f_v$ data in Fig. 11. Eq. 2 rederived with the expressions (21) for rod eutectics is:

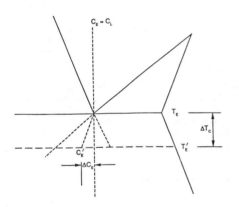

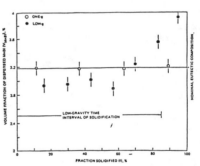

FIGURE 10
Hypothetical non-equilibrium phase diagram resulting for solid-liquid interface undercooling.

FIGURE 11
Volume fraction dispersed MnBi as a function of fraction sample length solidified, for MnBi/Bi eutectic samples solidified at 50 cm/hr in one-gravity and in low-gravity (13).

$$\lambda_{R1g}/\lambda_{R0g} = \left[ \frac{Z_{1g}^{\frac{1}{2}} (A_R + 1/f_{v1g})}{Z_{0g}^{\frac{1}{2}} (A_R + 1/f_{v0g})} \right]^{\frac{1}{2}} \tag{6}$$

Where $Z = f_v / (1-f_v)$. If we assume $A_R = 100$, and take $f_{v1g} = 0.032$ and $f_{v0g} = 0.028$ (Fig. 11), we obtain $\lambda_{R1g} = 1.05 \lambda_{R0g}$. This is only 10% of the

59

observed change.  This result is relatively insensitive to $A_R$
(e.g. for $A_R = 1000$, $\lambda_{Rlg}/\lambda_{ROg}=1.06$).  Thus,
the gravity dependent change in $f_v$, using the Jackson and
Hunt expression, can only partially explain low-g eutectic
spacings.

<u>$Fe_3C/Fe$:  Lamellar Eutectics</u>

$Fe_3C/Fe$ lamellar eutectic (Fig. 3) spacing decreases in
low-gravity periods by about 25%.  Clearly, predictions (17),
based on $Al_2Cu/Al$ results, that lamellar eutectic spacing is
not influenced by gravity are incorrect.  Greater dissymety in
liquidus slope ($m_\alpha/m_\beta = 2.5$ for $Fe_3C/Fe$ compared to 1.1
for $Al_2Cu/Al$) make $Fe_3C/Fe$ more sensitive to convective
influence on $f_v$ (Eq. 2).  The gradual gravity independent
decrease in the phase spacing (Fig. 3) indicates off-eutectic
type solidification, but whether or not interfacial undercooling
for $Fe_3C/Fe$ increases in low gravity has not yet been
established.  It is interesting to note that the eutectic grain
size for the stable Fe-C (iron-graphite) irregular off-eutectic
increases in low-gravity zones (12).

<u>Diffusion/Atomic Transport</u>

Neither on-eutectic nor off-eutectic convection models
predict the data in Table II.  Convective effects on thermal
gradient (27) and growth rate (28) also fail to explain the
low-gravity eutectic spacing.

Frohberg (29) suggests that gravity influences
solidification through microconvections driven by microscopic
concentration and temperature gradients or by thermodynamic
liquid density fluctuations.  These microconvections are
independent of the previously discussed macroconvection but are
indistinguishable from collective atomic motion - liquid
diffusion.

The effective diffusion coefficient, $D_{eff}$, for solidification
in normal gravity is:

$$D_{eff}(1g) = D + D_{wall} + D_{mic} + D_g \qquad (7)$$

where D is the intrinsic diffusion coefficient, and $D_{wall}$,
$D_{mic}$, and $D_g$ are contributions respectively of wall
effects, thermal and solutal microconvection, and thermodynamic
density fluctuations.  For zero gravity

$$D_{eff}(0g) = D + D_{wall} \leq D_{eff}(1g). \qquad (8)$$

Low-gravity liquid metal diffusion experiments for Zn (30)
and Sn (31) find $D_{eff}(0g)$ is less than $D_{eff}(1g)$ by
10-60%.  Stefanescu et. al. (12) suggest that a decrease in
$D_{eff}$ of this order results in the magnitude of lamellar
spacing decrease found for low-g solidified $Fe/Fe_3C$.  Similar
calculations for Bi/MnBi (32) were made using Magnin and Kurz
theory (33).  The shift in the locus of extrema (for rod
spacing versus undercooling) observed for low-g samples
relative to undamped one gravity samples was fit by a decrease
in $D_{eff}$ of 50%.

60

Equation 8 is not sufficient to explain systems (Al/Al$_3$Ni, Table II) where low gravity solidification increases eutectic spacing. Frohberg (29) suggests a diffusion boundary layer (stagnant film) approach. The thickness of the boundary layer during solidification (23) is $D_{eff}/R$. Macroconvection decreases the boundary layer thickness. This can approximately be described by lower $D_{eff}$. Al/Al$_3$Ni, it is hypothesized, has higher macroconvection in the ground experiments resulting in $D_{eff} < D$. The elimination of macroconvection in low gravity, then, increases $D_{eff}$.

This boundary layer explanation for larger eutectic spacing for solidification in low gravity has obvious problems. 1. For on-eutectic solidification, the solute redistribution occurs on the scale of the phase spacing not $D_{eff}/R$. 2. It is inconsistent to invoke large macroconvections during ground experiments only for Al/Al$_3$Ni, and not for InSb/NiSb (18) and Al/Al$_2$Cu (17) which were processed in similar apparatus and conditions. Thus, an adequate theory for the effect of natural convection on eutectic solidification remains to be developed.

## Conclusions

1. Low-gravity solidification of on-eutectic alloys results in an increase, decrease, or no change in phase spacings depending on the alloy system.

2. Numerical solution of the concentration fields at the solid-liquid interface for on-eutectic growth qualitatively predicts the trend of decreased interfacial undercooling and phase spacing with increased convection. However, the model predicts insignificant difference in phase spacing between alloy solidified in one-gravity from that in low-gravity.

3. The off-eutectic models predict positive, negative or zero change in spacing for low gravity solidification dependent on off-eutectic composition and phase diagram symmetry. The theory is modified to account for nonequilibrium phase diagram resulting from increased undercooling in low gravity. An expression is given for $\lambda_{1g}/\lambda_{0g}$ for rod eutectics. The model predicts only 10% of the change in $\lambda$ found in experiments for MnBi/Bi.

4. Unlike Al$_2$Cu/Al, Fe$_3$C/Fe, lamellar spacing is altered by low gravity solidification. Thus, contrary to previous suggestions, both lamellar and fibrous eutectics spacing can be sensitive to convection.

5. An adequate theory describing the effect of natural convection on eutectic spacing remains to be developed.

## Acknowledgments

This work was supported by the NASA Microgravity Science and Applications Office. Helpful discussions with Dr. R.J. Naumann, NASA/MSFC and Dr. W. F. Kaukler, UAH, during the preparation of this paper are gratefully acknowledged.

## References

1. M.C. Flemings, Solidification Processing (New York, N.Y.: McGraw-Hill, 1974).

2. M.E. Glicksman, E. Winsa, D. Hahn, T.A. Lograsso, E.R. Rubenstein, and E. Selleck, "Isothermal Dendritic Growth - A Low Gravity Experiment," Materials Processing in the Reduced Gravity Environment of Space ed. R.H. Doremus and P.C. Nordine (Materials Research Society, Pittsbusrgh, PA, 1987).

3. A.K. Sample and A. Hellawell, "The Mechanisms of Formation and Prevention of Channel Segregation During Alloy Solidification", Metallurgical Transactions, 15A (1984) 2163-2173.

4. S.M. Copley, A.F. Giamei, S.M. Johnson, M.F. Hornbecker, Metallurgical Transactions, 1A (1970) 2193.

5. J.L. DeCarlo and R.G. Pirich, "Effects of Applied Magnetic Fields During Directional Solidification of Eutectic Bi-Mn," Metallurgical Transactions, 15A (1984) 2155-21161.

6. D. Mattessen, "Low Gravity Crystal Growth Experiments," (Electronic Materials Group, MIT, Seminar presented to Space Science Laboratory, NASA, MSFC, 1986).

7. R.J. Naumann and H.W. Herring, Materials Processing in Space: Early Experiments, NASA Special Publication 443 (1980).

8. J.R. Carruthers, "Crystal Growth in a Low Gravity Envirnoment," Journal of Crystal Growth, 42 (1977) 379-385.

9. M.H. McCay, J.E. Lee, and P.A. Curreri, "The Effect of Gravity Level on the Average Primary Dendritic Spacing of a Directionally Solidified Superalloy," Metallurgical Transactions, 17A (1986) 2301-2303.

10. M.H. Johnston and R.A. Parr, "The Influence of Acceleration Forces on Dendritic Growth and Grain Structure," Metallurgical Transactions, 13B (1982) 85.

11. M.H. Johnston, P.A. Curreri, R.A. Parr, and W.S. Alter, "Superalloy Microstructural Variations Induced by Gravity Level During Directional Solidification," Metallurgical Transactions, 16A (1985) 1683-1687.

12. D.M. Stefanescu, P.A. Curreri, and M.R. Fiske, "Microstructural Variations Induced by Gravity Level during Directional Solidification of Near-Eutectic Iron-Carbon Type Alloys," Metallurgical Transactions, 17A (1986) 1121-1130.

13. D.J. Larson and R.G. Pirich, "Influence of Gravity Driven Convection on the Directional Solidification of Bi/MnBi

Eutectic Composites," Materials Processing in the
Reduced Gravity Environment of Space, ed. G.E.
Rindone (Materials Research Society, Pittsburgh, PA,
(1982) 523-532.

14. R.G. Pirich, D.J. Larson, and G. Busch, "Studies of
Plane-Front Solidification and Magnetic Properties of
Bi/MnBi,"AAIA Journal 19(5) (1981), 589-594.

15. L.L. Lacy and G.H. Otto, "Electrical Resistivity of
Galium-Bismuth Solidified in Free Fall," AIAA Journal,
13, (1975) 219.

16. M.K. Wu, J.R. Ashburn, P.A. Curreri, and W.F. Kaukler,
"Electrical Properties of Al-In-Sn Alloys Directionally
Solidified in High and Low Gravitational Fields,"
Metallurgical Transactions, 18A (1987) 1511-1517.

17. J.J. Favier and deGoer, "Directional Solidification of
Eutectic Alloys," Fifth European Symposium, Material
Sciences Under Microgravity, Results of Spacelab I,
ESA SP-222 (8-10 rue Mario-Nikis, 75738 Paris Cedex 15,
France, 1984) 127-133.

18. G. Muller and P. Kyr, "Directional Solidification of
InSb-NiSb Eutectic," Fifth European Symposium, Material
Sciences Under Microgravity, Results of Spacelab I,
ESA SP-222 (8-10 rue Mario-Nikis, 75738 Paris Cedex 15,
France, 1984) 141-146.

19. R.G. Pirch, G. Busch, W. Poit, and D.J. Larson, "The Bi-
MnBi Eutectic Region of the Bi-Mn Phase Diagram"
Metallurgical Transactions, 1 11A (1980) 193.

20. R.G. Pirich, "Spar IX Technical Report for Experiment 76-
22 Directional Solidification of Magnetic Composites,"
(Grumman Research and Development Center Report RE-642,
1982).

21. K.A. Jackson and J.D. Hunt, "Lamellar and Rod Eutectic
Growth", Transactions of Metallurgical Society of
AIME, 236 (1966) 1129-1142.

22. E.A. Hasemeyer, C.V. Lovoy, and L.L. Lacy, "Skylab
Experiment M566 Copper-Aluminum Eutectic," (NASA Report
M-74-5, Skylab Results, 1, (1971) 457-466.

23. J.D. Verhoeven and R.H. Homer, "The Growth of Off-Eutectic
Composites from Stirred Melts," Metallurgical
Transactions, 1 (1970) 33437-3441.

24. J.M. Quenisset and R. Naslain, "Effect of Forced
Convection on Eutectic Growth," Journal of Crystal
Growth 54 (1981) 465-474.

25. V. Baskaran and W.R. Wilcox, "Influence of Convection on
Lamellar Spacing of Eutectics," Journal of Crystal
Growth 67 (1984) 343-352.

26. G.F. Eisa, W.R. Wilcox, and G. Busch, "Effect of Convection on the Microstructure of MnBi/Bi Eutectic," Journal of Crystal Growth (in press).

27. P.S. Ravishankar, W.R. Wilcox and D. Larson, "The Microstructure of MnBi/Bi Eutectic Alloys," Acta Met. 28 (1980) 1583.

28. W.R. Wilcox, K. Doddi, M. Nair and D.J. Larson, "Influence of Freezing Rate Changes on MnBi/Bi Eutectic Microstructures," Advances in Space Research 3 (1983) 79.

29. G. Frohberg, "Diffusion and Atomic Transport," Materials Science in Space, eds. B. Feurbacher, H. Hamacher and R.J. Naumann (Springer Verlag, N.Y., 1985) 93-128.

30. A.O. Ukanwa, "Skylab Experiment M558," Proceedings of III Space Processing Symposium. Skylab Results (NASA/MSFC, Huntsville, AL (1979) 427.

31. G. Frohberg, K.H. Kraatz, and H. Wever, "Selfdiffusion of Sn112 and Sn124 in Liquid Tin," 5th European Symposium on Materials Science under Microgdravity. Results of Spacelab (Elmau, ESA, Paris, 1984) 201-205.

32. D.L. Larson, Jr., "The Influences of Convection on Directional Solidification of Eutectic Bi/MnBi," in preparation, 1987.

33. P. Magnin, "Competitive Stable/Metastable Solidification of Fe-C-X Eutectic Alloys," (Thesis No. 560, Ecole Polytechnique Federal de Lausanne, Switzerland, 1985).

DYNAMICAL STUDIES OF THE LAMELLAR EUTECTIC GROWTH

V. Seetharaman and R. Trivedi

Ames Laboratory, USDOE and the Department of
Materials Science and Engineering
Iowa State University
Ames, Iowa 50011

Abstract

Dynamical studies on the growth of the lamellar eutectic during directional solidification of a model, transparent, organic system have been
carried out. The evolution of the eutectic interface morphology and the
changes in the lamellar spacings have been followed as a function of time
after the growth velocity is changed suddenly from one constant value to
another. The kinetics of the change in lamellar spacing exhibit two regimes: regime 1 in which the spacing changes rapidly with time and regime 2
in which gradual adjustments in the lamellar spacing occur over long durations. When the velocity is increased such that the lamellar spacing is required to be reduced by a factor of two, this is found to be achieved through
catastrophic shape instabilities followed by either the nucleation of β
crystals at the center of the α lamellae or by the splitting of the β
lamellae. However, when the extent of decrease in the lamellar spacing is
much larger, a new mechanism for spacing reduction has been identified.
This mechanism involves the degeneracy of the initial cooperative growth
morphology through the formation of a thin, continuous layer of a single
phase solid and the subsequent renucleation and growth of the α/β lamellae.
An increase in the lamellar spacing is accomplished by the preferential
elimination of the lamellae with smaller local spacings. The mechanisms
responsible for the rapid and the gradual changes in the lamellar spacings
observed in the different regimes are discussed.

Solidification Processing of Eutectic Alloys
D.M. Stefanescu, G.J. Abbaschian and R.J. Bayuzick
The Metallurgical Society, 1988

## Introduction

The steady state growth of lamellar eutectics has been studied extensively by both the theoretical and experimental investigators. Jackson and
Hunt [1] have developed a detailed model to predict the relationships
between the lamellar spacing, $\lambda$, interface undercooling, $\Delta T$ and the growth
velocity, V. This model also permits the calculation of interface shapes
and their response to small fluctuations in the growth rate. It is interesting to note that this model as well as the more recent models proposed by
Langer and his coworkers [2-4] and by Cline [5-6] predict the existence of a
wide range of lamellar spacings for any given growth velocity under steady
state growth conditions. However, the bulk of experimental studies on
eutectic growth has shown that the range of steady state lamellar spacings
is very narrow and that the average spacing lies close to that corresponding
to minimum undercooling [7].

In contrast to the steady state growth, the dynamical adjustments in
the eutectic morphology and the lamellar spacing occurring in response to
the variations in the growth velocity have not been studied very extensively. Jackson and Hunt [1] have studied the changes in interface morphology to imposed velocity changes in very thin cells of the transparent
organic eutectic of carbon tetrabromide and hexachloroethane. They have
shown that shape instabilities in the wider phase of the eutectic lead to
catastrophic changes in the lamellar spacing. Boettinger [8] has carefully
investigated the changes in lamellar spacings following a rapid increase,
decrease or cycling of the growth velocity in thick films of Pb-Sn-Cd
ternary eutectic. He has shown that when the final interface velocity is
greater than four times the initial velocity, the lamellar spacing does not
decrease uniformly with time but decreases in stages to its final value. He
has also concluded that the overall change in the lamellar spacing following
a change in interface velocity from one fixed value to another does not
restore the value of $\lambda^2 V$ to its original value. Carlberg and Fredriksson
[9] have investigated the kinetics and mechanisms of changes in the lamellar
spacing occurring in the directionally solidified bulk samples of Ag-Cu,
Al-Cu and Al-Zn alloys when the growth velocity was increased continuously.
While the Ag-Cu system exhibits step changes in the lamellar spacing, the
Al-Zn and Al-Cu systems tend to adjust their lamellar spacings in a
continuous manner.

The aim of the present investigation is to study the kinetics and
follow the mechanisms of lamellar spacing changes occurring under dynamical
conditions of eutectic growth. The transparent eutectic system of carbon
tetrabromide-hexachloroethane has been chosen for this study for the following reasons: (i) in-situ observations of the adjustments in lamellar
spacings at the solid-liquid interface are possible (ii) the value of $\lambda^2 V$
for this system is nearly an order of magnitude larger than the corresponding values for the metallic eutectics. This improves the precision with
which lamellar spacings can be measured and (iii) a careful study of the
steady state growth behavior of this eutectic system has been carried out by
the authors [10]. This paper describes the evolution in the interface
morphologies and the changes in the lamellar spacings occurring after a
sudden increase or decrease in the growth velocity in this system. A qualitative description of the mechanisms responsible for the observed morphological changes is also offered.

## Experimental Procedure

Carbon tetrabromide ($CBr_4$) and hexachloroethane ($C_2Cl_6$) were purified
by sublimation at suitable pressures and temperatures. Alloys corresponding

66

to the eutectic composition of 8.4 wt% hexachloroethane were melted and
filled in thin glass cells of 150 μm thickness. Directional solidification
experiments were carried out on these alloys using an apparatus described in
detail by Mason and Eshelman [11]. A constant temperature gradient of 3.6
K/mm was maintained for all the experiments described in this paper. Additional details on the purification of materials, alloy preparation and the
directional solidification runs have been described elsewhere [10]. The
following three different types of experiments were conducted to study the
dynamical changes in the morphology and the lamellar spacing occurring at
the eutectic interface: (i) The solid/liquid interface was initially held
stationary ($V_i = 0$) for one hour and then the velocity was increased rapidly
to a specific value, V < 2.0 μm/s and held at that value till steady state
conditions were attained. (ii) The sample was initially solidified at a
low velocity for a sufficient time to achieve a steady-state distribution of
spacings. The velocity was then quickly increased to a new value, V such
that $V_i$ < V < 2.0 μm/s and held at this value till steady state was achieved. (iii) The sample was initially solidified at a high velocity and then
the velocity was decreased suddenly to a new value V such that V < $V_i$ < 2.0
μm/s and held at this velocity for a sufficient time to achieve steady-state
conditions.

## Results

### Evolution of Eutectic Microstructures

In order to compare the dynamics of eutectic spacing changes occurring
in various velocity change experiments, it is first necessary to characterize steady-state eutectic spacings. A detailed study on the effect of
velocity on steady-state eutectic spacings has been carried out by the
authors [10] in the $CBr_4$-$C_2Cl_6$ system. In this study, the evolution of
steady-state spacing was examined when the velocity was increased from zero
to the desired value. Since these results are germane to the experimental
study presented in this paper, a brief description of these results will be
presented in this section.

Fig. 1 illustrates the typical nucleation and growth processes observed
at a growth velocity of 0.5 μm/s at different times. When the cell was
initially held stationary, the solid-liquid interface was found to be planar
and the solid consisted of a single phase (α) structure. When the velocity
was increased to 0.5 μm/s, the single phase interface persisted for a few
minutes until the interface undercooling increased sufficiently to nucleate
the eutectic lamellae. Initially, the nucleation commenced at one edge of
the cell and then the fine lamellar structure propagated very rapidly along
the solid-liquid interface towards the other edge [Fig. 1(a)]. Although the
eutectic structure was rather ill-defined at that stage, it quickly developed into a periodic array [Fig. 1(b)]. Subsequently, some of these lamellae
were eliminated and this led to a significant increase in the interlamellar
spacing with time [Fig. 1(c) and 1(d)]. This elimination process occurred
quite randomly and led to a continuous increase in the average interlamellar
spacing with time. The different stages of nucleation and growth described
above were typical and observed at all growth velocities in the range 0.1 to
2.0 μm/s. Fig. 2 shows the variations in the interlamellar spacings with
time for different growth velocities. The main features of this plot are:
(1) The time required for the nucleation of the eutectic decreases sharply
with the velocity; (2) The initial rate of coarsening of the lamellar is
very rapid at all velocities; thereafter, the process of elimination of the
lamellae becomes very sluggish reading to a gradual increase in the interlamellar spacing with time until steady state conditions are established.
Let $\tau_1$ and $\tau_2$ be the times required for the onset of nucleation of the

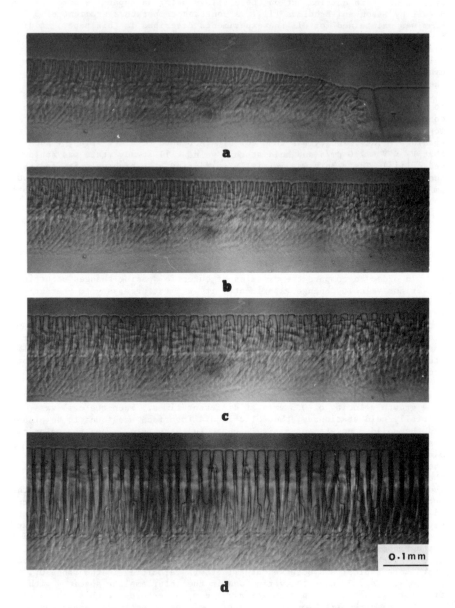

0.1mm

Fig. 1. Evolution of lamellar structures during the directional solidification of the $CBr_4$-8.4 wt% $C_2Cl_6$ eutectic alloy at a velocity of 0.5 µm/s. The times measured from the instant at which the velocity was changed from zero to 0.5 µm/s are (a) 11.33 min. (b) 11.67 min. (c) 12.33 min. and (d) 16.9 min.

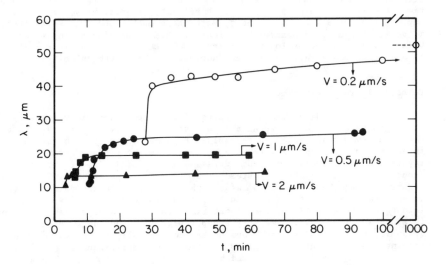

Fig. 2.  Time evolution of the mean interlamellar spacings during the
solidification of the eutectic at different velocities.

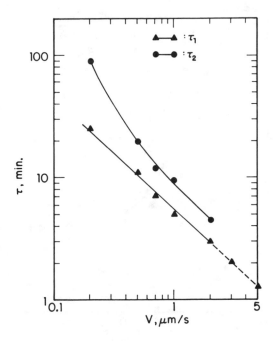

Fig. 3.  Variations of the time constants, $\tau_1$ and $\tau_2$ for lamellar growth
with the growth velocity.

eutectic structure and for the attainment of a lamellar spacing equal to 0.9
times the steady state spacing at each velocity, respectively. Figure 3
depicts the variations in $\tau_1$ and $\tau_2$ with the growth velocity. While the log
$\tau_1$ vs log V plot is linear with a slope of -0.92, the plot of log $\tau_2$ vs log
V is slightly curved at low velocities. Nevertheless, it is clear that both
$\tau_1$ and $\tau_2$ decrease rapidly with an increase in the growth velocity.

## Velocity Increase Experiments

This section describes the results of the experiments in which steady
state growth conditions were established at an initial velocity, $V_i$ and then
the velocity was increased quite rapidly to a new value, V. The average
interlamellar spacings, $\lambda$ were measured at different times after the velocity was changed from $V_i$ to V and expressed as a ratio, $\lambda/\lambda_i$ of the instanta-
neous lamellar spacing to the steady state lamellar spacing, $\lambda_i$ at $V_i$.
Figure 4 shows the variation of $\lambda/\lambda_i$ with time for different values of $V/V_i$.
These data show that the change in $\lambda/\lambda_i$ with time can be broadly divided
into three stages: (1) At very short time, there is no appreciable change
in spacing. (2) $\lambda$ decreases very steeply with time and (3) $\lambda$ decreases very
gradually at long durations. Thus we can conceptually divide the change in
spacing with time into two regimes: Regime 1 characterized by a rapid fall
in $\lambda/\lambda_i$ and Regime 2 characterized by a slow decrease in $\lambda/\lambda_i$. The mechanisms responsible for the changes in lamellar spacing in these two regimes
are quite different and will now be described.

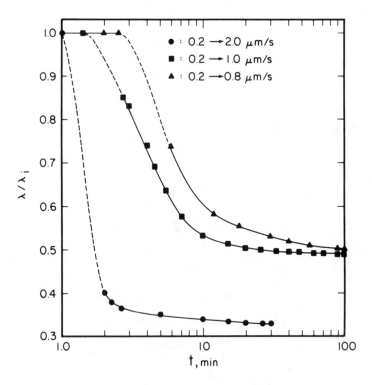

Fig. 4.  Time dependence of the reduction in the lamellar spacing ratio $\lambda/\lambda_i$
observed in the velocity increase experiments.

The morphological changes of the interface occurring in regime 1 can be broadly classified into two types:

(i) those occurring when the spacing decreases by a factor of two, i.e., when $V/V_i \approx 4$.

(ii) those occurring when the ratio of the initial spacing to the final spacing is not an even integer i. e., $V/V_i \neq 4, 16....$

Even for the special case, $V/V_i = 4$, two different mechanisms of spacing reduction have been observed and these will be described first. This will be followed by the description of the mechanisms operating in the more general case.

Figure 5 reveals the early stages in the microstructural evolution when the solidification velocity was increased from 0.2 to 0.8 μm/s. The transition from a flat α/liquid interface to a concave α/liquid interface, Fig. 5(a) and 5(b), widening of the β lamallae, development of concavity in the β lamellae and the formation of ridges along the α/liquid interface at equal distances from the α/β boundaried [Fig. 5(c)] are seen clearly. Figure 5(d) and 5(e) reveal that the ridges along the α/liquid interface mentioned above, ultimately become the sites for the formation of thin β lamellae. This leads to the development of a periodic array of α and β lamellae [Fig. 5(f)] with interlamellar spacing approximately one half of that obtained at $V_i = 0.2$ μm/s. It is interesting to note the apparent continuity of the β lamellae which suggests that the β lamellae have possibly undergone fork shaped symmetric branching.

An alternate mechanism, proposed originally by Jackson and Hunt [1] for the reduction in the lamellar spacing by a factor of two is evident in Fig. 6. This sequence of micrographs illustrates the changes in the interface morphology when the solidification velocity is increased from 0.5 to 2.0 μm/s. The different stages involved in the microstructural response to the increase in velocity are:

(i) development of a concave α/liquid interface [Fig. 6(a) and (b)]

(ii) nucleation of β crystals in these concave regions in the center of the α lamellae [Fig. 6(c) and (d)] and

(iii) growth of these freshly nucleated β lamellae and the thinning of original β lamellae [Fig. 6(e) and (f)].

Again, the final microstructure consists of a uniform array of α/β lamellae with an average spacing equal to approximately one half of that of the initial array.

The microstructural changes occurring after the velocity is increased from 0.2 to 2.0 μm/s are illustrated in Fig. 7. In this general case where the spacing does not decrease by a factor of two, the reduction in spacing can not be accomplished by a simple splitting of the α or β phase. Instead, it involves a more complex sequence of events as shown in Fig. 7.

The different mechanisms described above and illustrated in Figures 5-7 represent the events occurring during the rapid initial transient (regime 1). During the time represented by the regime 2, fine-tuning of the interlamellar spacing occurs principally by the formation and movement of the lamellar faults. Figure 8 shows an example of a lamellar fault occurring in the longitudinal section of the $CBr_4-C_2Cl_6$ eutectic growing at $V = 1$ μm/s. The mechanism by which the migration of the faults enables adjustments in the lamellar spacing as well as the conditions which favor this process are discussed subsequently.

71

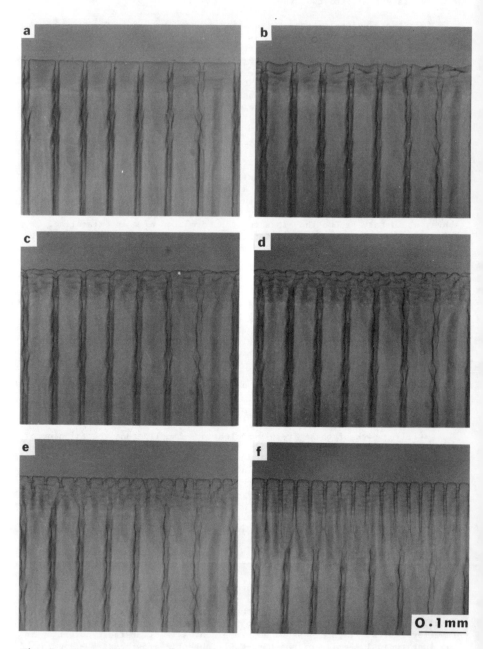

Fig. 5. Changes in the interface morphology after the velocity is increased from $V_i = 0.2$ µm/s to $V = 0.8$ µm/s (a) 15 s (b) 165 s (c) 195 s (d) 225 s (e) 270 s and (f) 420 s.

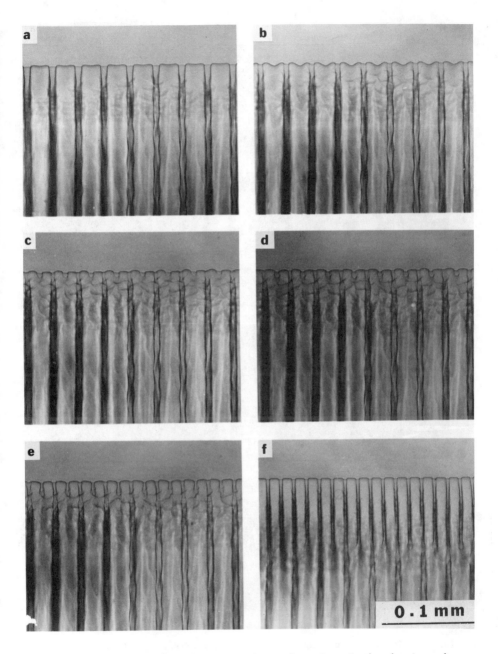

Fig. 6.  Changes in the interface morphology after the velocity is changed from $V_i$ = 0.5 μm/s to V = 2.0 μm/s (a) 30 s (b) 70 s (c) 85 s (d) 90 s (e) 100 s and (f) 310 s.

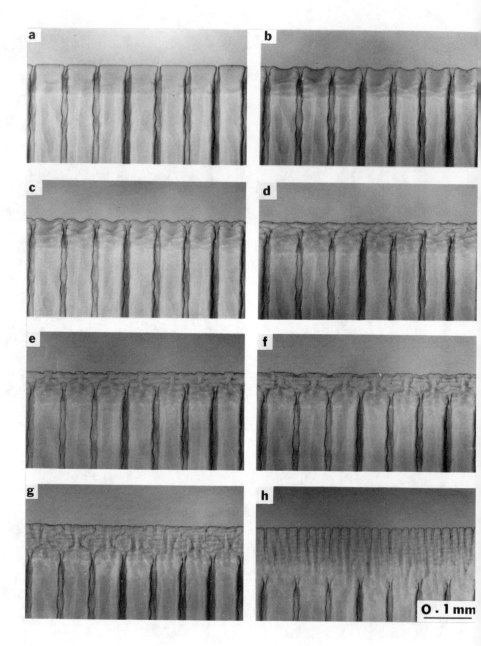

Fig. 7. Changes in the interface morphology after the velocity is increased from $V_i = 0.2$ μm/s to $V = 2.0$ μm/s (a) 20 s (b) 40 s (c) 60 s (d) 70 s (e) 75 s (f) 85 s (g) 100 s and (h) 140 s.

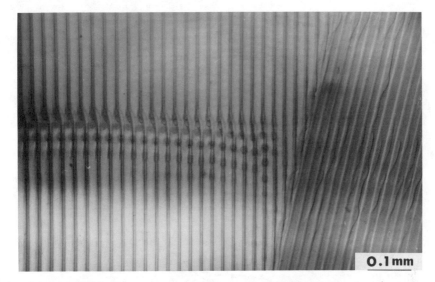

Fig. 8  The lamellar structure of the eutectic growing at $V = 1$ μm/s
($V_i = 0.2$ μm/s) for 45 minutes.  The structure reveals the presence
of a mismatch boundary as seen in a longitudinal section.


An entirely new mechanism for the reduction in the average lamellar
spacing has been observed in a few experiments in which $V/V_i = 4$.  In this
case, the lamellae move gradually towards the centre of the sample and grow
at an angle of ~3-6° to the heat flow direction.  Both the α and β lamellae
undergo a reduction in thickness and this leads to a net decrease in the
lamellar spacing.  This mechanism envisages the nucleation of lamellae at or
near the cell edges or grain boundaries and their lateral displacement
towards the center of the cell.


## Velocity Decrease Experiments

This section deals with the experiments in which the solidification
velocity was decreased suddenly from an initial velocity, $V_i$ to a new
velocity, $V$.  Figure 9 summarizes the variation of $\lambda/\lambda_i$ with time for
different values of $V/V_i$.  The plots of $\lambda/\lambda_i$ versus t consist of two
regimes:  i) regime 1 in which $\lambda/\lambda_i$ increases very steeply with time and
(ii) regime 2 in which $\lambda/\lambda_i$ increases very gradually.  It should be noted
that the final values of $\lambda/\lambda_i$ obtained after fairly long durations at $V =$
0.2 μm/s are clearly well below the expected values of $\lambda/\lambda_i$ based on the
relation $\lambda^2 V =$ constant.  This suggests that the processes leading to an
increase in $\lambda/\lambda_i$ upon a decrease in the solidification velocity are quite
sluggish.

The initial and final stages of the microstructural changes occurring
as a result of the decrease in the velocity from $V_i = 2$ μm/s to $V = 0.2$ μm/s
are illustrated in Fig. 10 and 11, respectively.  Figure 10 clearly demonstrates the termination of both α and β lamellae.  When a β lamella gets
eliminated, the adjacent α lamallae join together to form a single α lamella
which generally reduces its thickness in relation to the thickness of the

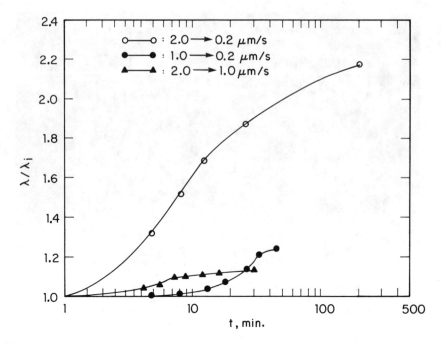

Fig. 9 Time dependence of the increase in the lamellar spacing ratio, $\lambda/\lambda_i$ observed in the course of the velocity decrease experiments.

adjoining β lamellae. In contrast, when an α lamella gets eliminated, the bounding β lamellae join together to produce a single β lamella which gradually increases its thickness in order to maintain the ratio of the volume fractions of the α and β phases at a constant value. These termination events occur quite randomly along the interface. The bulk of the terminations are completed within 30 minutes [Fig. 10(c)] thereafter, the eliminations occur only very occasionally [Fig. 11(a)] and at those locations at which the local lamellar spacing is smaller than the average lamellar spacing of the entire array. Even after 200 minutes at V = 0.2 μm/s, $\lambda/\lambda_i$ has reached a value of only 2.2 compared to the expected value of 3.16. At this stage, the lamellae have organized themselves into a nearly uniform array and therefore additional elimination events are very unlikely to occur. Any further increase in spacing can be achieved only by the formation and migration of the lamellar faults. However, the lamellar spacing at this stage is approximately one-fifth of the sample cell thickness. In view of this, the nucleation and propagation of faults become very unfavorable. Thus, the lamellar spacing ratio, $\lambda/\lambda_i$ will tend to saturate at this value of 2.2 for the limited sample thickness used in this study.

## Discussion

### Kinetics of Lamellar Spacing Adjustments

If the solidification velocity is changed suddenly, it causes the interlamellar spacings to adjust such that the new spacing becomes optimum

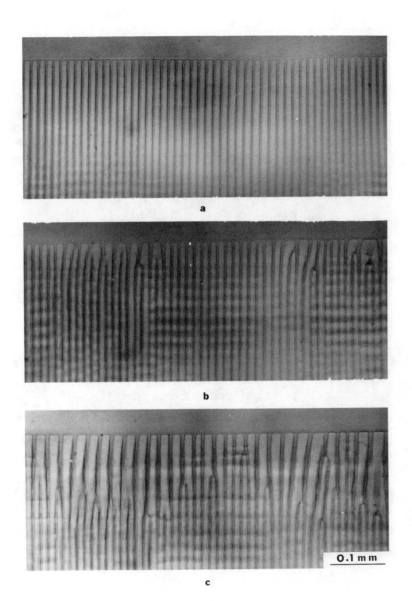

Fig. 10. Initial stages of the increase in the lamellar spacing through the elimination of the lamellae ($V_i$ = 2.0 µm/s, V = 0.2 µm/s) (a) 10 s (b) 290 s and (c) 740 s.

O.1 mm

Fig. 11.   Final stages of the coarsening of the lamellae ($V_i$ = 2.0 µm/s and V = 0.2 µm/s) (a) 27 min (b) 200 min.

78

with respect to solute diffusion and surface energy considerations. The kinetics of such spacing adjustments resulting from an increase or decrease in the velocity exhibit some similarities and some dissimilarities. Both are characterized by an initial rapid transient (regime 1) and a gradual change at long durations (regime 2). However, the duration of regime 1 for the velocity increase experiments is much shorter than that for the velocity decrease experiments. Let us designate $\tau_3$ and $\tau_4$ as the characteristic times corresponding to the inflection points in the $\lambda/\lambda_i$ versus t plots for the velocity increase (Fig. 4) and for the velocity decrease experiments (Fig. 9), respectively. Table I lists the values of $\tau_3$ and $\tau_4$ for different values of $V/V_i$ or $V_i/V$. It is clear that both $\tau_3$ and $\tau_4$ increase sharply with the decrease in $V/V_i$ or $V_i/V$. At the same time it is important to note that $\tau_4$ is nearly 6-8 times longer than $\tau_3$ for the same values of $V/V_i$ for the velocity increase experiments and $V_i/V$ for the velocity decrease experiments. These differences in the values of $\tau_3$ and $\tau_4$ can be directly attributed to the fact that final velocities differ considerably. If experiments were conducted at constant values of both $V/V_i$ (or $V_i/V$) and V then the values of $\tau_3$ and $\tau_4$ will be comparable.

Table I
Time constants for the velocity
change experiments

| $\dfrac{V}{V_i}$ or $\dfrac{V_i}{V}$ | $\tau_3$ (min) | $\tau_4$ (min) |
|:---:|:---:|:---:|
| 10 | 1.6 | 10 |
| 5 | 3.2 | 24 |
| 4 | 4.5 | 30 |

## Mechanisms for the Catastrophic Multiplication of Lamellae

When the solidification velocity is increased suddenly, the original lamellar spacing is too large to be stable. The increase in velocity leads to the piling up of the excess solute in front of the wider phase ($\alpha$) which, in turn, lowers the freezing temperature locally. This renders the central regions of the $\alpha$ lamellae concave. This is indeed observed as a common feature among the different sequences of the changes in the interface morphology illustrated in Fig. 5-7. Similarly, a concavity may also develop in the center of the $\beta$ lamellae and then the $\beta$ lamellae will tend to branch. This is followed by the nucleation of the $\alpha$ phase crystals at the intervening space between the two branches of the $\beta$ lamellae. The micrograph shown in Fig. 5(c) appears to provide evidence for these steps, which are represented schematically in Fig. 12(a). Although the final microstructure shown in Fig. 5(f) seems to indicate some kind of continuity between the original and the new $\beta$ lamellae, the microstructures corresponding to the intermediate stages [Fig. 5(d) and 5(e)] do not have sufficient contrast to provide conclusive evidence for the continuous growth of the branched $\beta$ lamellae. In fact, a detailed examination of these and other intermediate micrographs suggests that the initially split $\beta$ lamellae have ceased to grow and that the $\alpha$ crystals, formed between these branches, continue to widen till a thin continuous layer of $\alpha$ phase, composed of crystals of different orientations and compositions, forms at the interface. Subsequently, $\beta$ crystals may renucleate along the new interface and this would lead to the cooperative growth of the $\alpha/\beta$ lamellae with a lamellar spacing approximately one half of that of the original array.

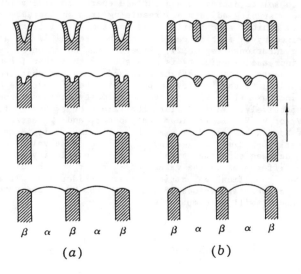

$$\beta \quad \alpha \quad \beta \quad \alpha \quad \beta \qquad \beta \quad \alpha \quad \beta \quad \alpha \quad \beta$$

$$(a) \qquad\qquad\qquad (b)$$

Fig. 12   Schematic diagrams showing the changes in the interface shape
         leading to a reduction in spacing following a sudden increase in
         the growth rate [12].   The arrow indicates the direction in which
         the changes occur with increasing time.

An alternate mechanism for the reduction in the lamellar spacing by a
factor of two is shown in Fig. 6.   It involves the nucleation and growth of
the β phase in the solute rich pockets associated with the concave regions
of the α lamellae.   Fig. 12(b) depicts schematically the different steps
involved in this process.   The micrographs shown in Fig. 6 provide a conclusive evidence for the operation of this mechanism, suggested originally by
Jackson and Hunt [1].

It is important to recognize the basic difference between the two
lamellar multiplication mechanisms observed in Figs. 5 and 6, although both
these mechanisms lead to a net reduction in the lamellar spacing by a factor
of two.   The mechanism observed in Fig. 6 results in the formation of the
new β lamellae in the center of the original α lamellae.   In contrast, the
mechanism observed in Fig. 5 shows the formation of the new β lamellae which
are displaced on either side of the center of the original α lamellae by
about one fourth the width of the original α lamellae.   This difference in
the relative positions of the original and the new lamellae suggests very
clearly the operation of two entirely different bifurcation modes for the
eutectic lamellae.   The choice of a specific bifurcation mode which operates
under given conditions depends on several factors.   For example, in the $CBr_4$
– $C_2Cl_6$ eutectic system, the extent of solute enrichment required for the
nucleation of β is much higher than that required for the nucleation of α.
This should promote the operation of the mechanism involving the initial
branching of β and the nucleation of α at relatively low velocities in this
system.   The other mode involving the nucleation of β should be favored at
higher velocities.   The observation of the latter mechanism for $V_i$ = 0.5 and
V = 2.0 μm/s and the plausible operation of the former mechanism for $V_i$ =
0.2 and V = 0.8 μm/s are in agreement with the above discussion.

80

The mechanism for the spacing reduction which occurs when $V/V_i$ is significantly larger than four, will now be examined. Figure 7 illustrates the sequence of events occurring after the velocity has been increased from 0.2 to 2.0 μm/s, i.e. $V/V_i$ = 10. The widening of the β lamellae in the form of a funnel [Figs. 7(c) and 7(d)] and the apparent continuity of the α/β boundaries seen in Figs. 7(g) and 7(h) seem to suggest the successive operation of the scheme sketched in Fig. 12(a). However, the appearance of a thin single phase layer just below the solid-liquid interface and the associated instabilities of this interface at the intermediate stages, Figs. 7(e) and 7(f), suggest that the operative mechanism is very analogous to that observed in Fig. 5. The weak contrast associated with the α/β boundaries near the solid-liquid interface as well as the inadequate spatial resolution inherent in the dynamical experiments do not permit precise description of the interface morphology at the intermediate times. However, the general mechanism of spacing reduction can be described as a two-step process: (i) the initial change by the mechanism outlined in Fig. 12(a) or 12(b) and (ii) the subsequent cessation of the growth of the β lamellae leading to the formation of a thin continuous layer of the α-phase at the solid-liquid interface, development of instabilities along the planar α/liquid interface and the final nucleation and growth of the α/β lamellar structure.

The change in spacing by the step (i) has been observed by Boettinger [8] in his experiments on Pb-Sn-Cd ternary eutectic in which the interface velocity was increased by changing the liquid temperature gradient, and by Carlberg and Fredriksson [9] in their continuous acceleration experiments on the Ag-Cu eutectic. It is, however, important to realize that evidences for the operation of the mechanism described in step (ii) can be obtained only through in-situ experiments. The transparent nature of our system has enabled us to observe the mechanisms described in steps (i) and (ii) in-situ.

Fine Adjustment of Lamellar Spacings

The catastrophic instabilities leading to a rapid alteration in the lamellar spacings in regime 1 are followed by very gradual changes in or fine tuning of the lamellar spacings in regime 2. Such fine adjustments in spacings can be brought about by the generation and migration of faults through the eutectic lamellae [1, 13, 14]. The origin of these lamellar faults has been examined by Double [15] and by Dean and Gruzleski [14]. Initially, a ripple-like instability appears in a group of two or three lamellae; as solidification proceeds, the misorientation spreads from one lamella to another till a discrete mismatch boundary is formed over 10-20 lamellae. The lattice misorientaiton introduced by the formation of such a sub-boundary may be accommodated by the simultaneous creation of another sub-boundary, presumably of opposite sign, at a distance of about ten lamellar spacings away.

A transverse section of a directionally solidified Pb-23.73 wt% Cd eutectic is shown in Fig. 13. This figure illustrates the two types of faults which occur commonly in lamellar eutectics: (i) the no-net fault (no-net mismatch boundary) which has equal number of lamellae on either side of the fault boundary, Fig. 13(a) and (ii) the net fault which is associated with an extra lamella on one side of the fault boundary, Fig. 13(b). Experimental results presented in Fig. 4 show that the rate at which the fine adjustment in spacing occurs is quite small. Also, the spacing does not decrease sufficiently to match with the steady-state value corresponding to a given final velocity. Thus, two aspects of faults need to be examined: (1) the ability of the fault to change the spacing, and (2) the factors which control the density of faults.

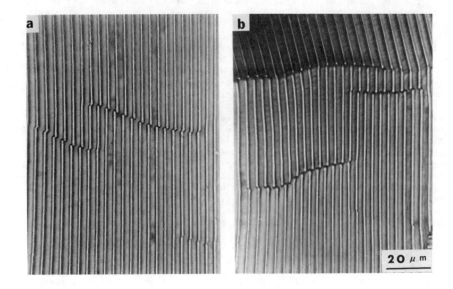

Fig. 13   Transverse section of the Pb-23.73 wt% Cd eutectic grown at 2.38
μm/s showing (a) no-net faults and (b) net faults.


        Both no-net faults and net faults contribute to the spacing change,
although net faults are more mobile than the no-net faults [14]. The no-net
faults can contribute to the spacing adjustment by a mechanism which involves their swiveling around the growth axis [15], as sketched in Fig. 14.
Even though the net number of lamellae above and below the boundary is the
same, the distribution of the projected lamellar spacing at the growth front
is different. A localized compression of spacings is seen in regions marked
A and A' whereas a localized expansion is noticed in regions marked B and
B'. When the solidification velocity is increased, a reduction in lamellar
spacing can be realized if the regions A and A' outgrow regions B and B'.
This is achieved by a swiveling movement of the boundary into an S-shaped
configuration. This counter-clockwise movement, taken to its conclusion,
would result in the alignment of the fault parallel to the lamellae and
thereby produce a fault-free grain with reduced spacing. Evidently, a
clockwise movement will produce an overall increase in lamellar spacing. It
should also be noted that a no-net fault can transform into a net fault or
vice versa by combining with a lamellar termination of proper sign.

        Jackson and Hunt [1] have proposed that the passage of single net
faults or terminations is responsible for adjustments in lamellar spacing.
However, Double [14] has demonstrated that terminations are generally im-
mobile and unimportant whereas the mismatch boundaries with their enhanced
mobility are crucial in producing the spacing changes. The mismatch boundaries move rapidly perpendicular to the lamellae such that those subgrains
with spacings closest to the optimum value grow at the expense of others.
Although the individual fault boundaries move only a small distance, the
presence and movement of a high density of faults produces a cumulative
effect in altering the lamellar spacings. Thus, it is important to examine
the factors which control the density of faults.

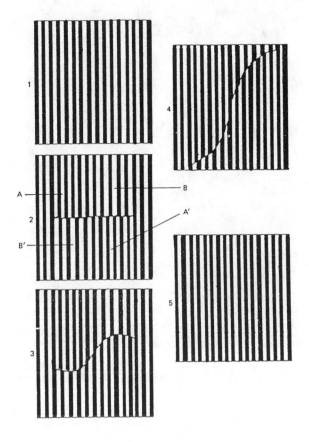

Fig. 14   A mechanism for the lamellar spacing adjustments by the swiveling
          motion of the mismatch boundary [15].


     The ease with which the mismatch boundaries can be created within the
eutectic grains depends on the grain size normal to the growth direction.
Fig. 15, which shows the transverse section of the Pb-23.73 wt% Cd alloy,
clearly brings out the variations in the density of the faults from one
grain to the other.  Since faults are responsible for effecting finer spacing adjustments, it is not surprising to find a good correlation between the
density of faults and the lamellar spacing in this figure.  It is evident
that fine grains with a low density of faults have larger interlamellar
spacings than coarse grains with a high density of faults.  This observation
is important in understanding the role of lamellar faults in the adjustment
of spacings in the $CBr_4$-$C_2Cl_6$ system.  Since the thickness of the solidifying mass in our experiment is only about 150 μm, it will be difficult to
form lamellar faults if the average interlamellar spacing is larger than
~20-25 μm.  Even for smaller spacings, the density of faults would be quite
low.  This explains the observation of fault free eutectic structures in
this system at low growth rates.

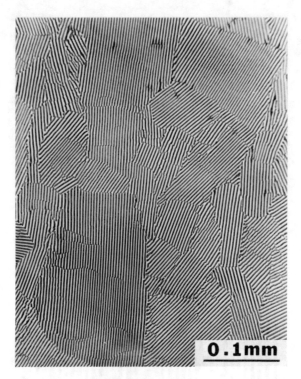

**0.1mm**

Fig. 15   Variations in the fault density and the interlamellar spacing with
the eutectic grain size in the Pb-23.73 wt% Cd eutectic.


In addition to the grain size, the density of faults depends on some
system parameters.  Carlberg and Fredriksson [9] have analyzed the
solidification behavior in three eutectic system viz., Ag-Cu, Al-Zn, and
Al-Cu and have correlated the ease of spacing adjustments in these systems
with their physical constants through their effect on the nature of the
solidification front.  The undercooling, $\Delta T$ at the eutectic interface is
related to the solidification velocity, V and the interlamellar spacing, $\lambda$
as follows [1]:

$$\Delta T = K_1 \ V \ \lambda \ + \ K_2/\lambda$$

where $K_1$ and $K_2$ are material dependent constants.  Carlberg and Fredriksson
[9] have shown that the eutectic interface will, in general, be rough due to
the distribution in lamellar spacings at any given velocity.  This roughness
is, however, dependent on the value of the ratio, $K_2/K_1$.  With an increase
in the value of $K_2/K_1$, from $3.0 \times 10^{-8}$ $mm^3/s$ for Al-Zn to $8.0 \times 10^{-8}$ $mm^3/s$
for Ag-Cu, the solidification interface has been shown to have become
smoother.  At the same time, they have demonstrated that the mechanism for
the reduction in lamellar spacings changes from the migration of mismatch
boundaries in the Al-Zn alloy (gradual increase in spacing with time) to the
catastrophic instability associated with lamellar multiplication in the
Ag-Cu system (step changes in spacing with time).  Careful experiments conducted by the present authors [10] on the steady state growth of the

eutectic lamellae in the $CBr_4$ - $C_2Cl_6$ system have yielded a value of $K_2/K_1$ = $2.67 \times 10^{-7}$ mm$^3$/s. Since this value is nearly three times that corresponding to the Ag-Cu system, it is logical to expect the enhanced role of catastrophic instabilities and the diminished role of the mismatch boundaries in the $CBr_4$ - $C_2Cl_6$ system. The experimental observations reported in the current work are consistent with this analysis.

## Conclusions

The following major conclusions can be drawn form this dynamical study on the kinetics of and the mechanisms for the changes in lamellar spacings in the $CBr_4$ - $C_2Cl_6$ eutectic system: (i) When the solidification velocity is increased from zero to any value, V < 2.0 μm/s, the eutectic structure forms by the simultaneous nucleation and growth of very fine crystals of α and β at the eutectic interface. This fine admixture of α and β crystals grows into an orderly arrangement consisting of alternate lamellae and then coarsens steadily to yield a characteristic steady state lamellar spacing. (ii) The kinetics of the change in lamellar spacings observed after an increase or decrease in velocity in the range $0.2 < V < 2.0$ μm/s exhibit two regimes: a rapid initial change (regime 1) and a gradual change at long durations (regime 2). The overall kinetics of the reduction in the lamellar spacing is found to be faster than the kinetics of coarsening for a given pair of velocities. (iii) When $V/V_i \approx 4$, the reduction in lamellar spacing is achieved through catastrophic shape instabilities involving either nucleation of β crystals at the center of the α lamellae or the branching of the β lamellae. (iv) When $V/V_i = 10$, a reduction in the lamellar spacing is achieved through a sequence of steps involving degeneracy of the initial cooperative growth morphology by the formation of a thin layer of perturbed single phase solid and the subsequent renucleation and growth of the α/β lamellae. (v) The increase in spacing associated with a decrease in velocity is achieved by the continuous elimination of the lamellae. These elimination events are controlled by the values of the local lamellar spacings. (vi) The role of lamellar faults in causing gradual changes in the lamellar spacing in this system is very limited. This is due to the difficulty in the creation and movement of faults/mismatch boundaries when the ratio of the sample thickness to the mean lamellar spacing is not sufficiently large.

## Acknowledgments

The authors thank Professor John Verhoeven for many helpful discussions and also for providing his unpublished results on the Pb-Cd Alloy. This work has been carried out at Ames Laboratory which is operated for the U. S. Department of Energy by Iowa State University under contract no. W-7405-ENG-82 and has been supported by the Office of Basic Energy Sciences, Division of Materials Sciences.

References

1. K. A. Jackson and J. D. Hunt, "Lamellar and Rod Eutectic Growth", Transactions of the Metallurgical Society of AIME, 236 (1966) 1129-42.

2. J. S. Langer, "Eutectic Solidification and Marginal Stability", Physical Review Letters, 44 (1980) 1023-26.

3. V. Datye and J. S. Langer, "Stability of Thin Lamellar Eutectic Growth", Physical Review B, 24 (1981) 4155-69.

4. V. Datye, R. Mathur and J. S. Langer, "Mode Selection in a Caricature of Eutectic Solidification", Journal of Statistical Physics, 29 (1982) 1-16.

5. H. E. Cline, "Growth of Eutectic Thin Films Structures", Journal of Applied Physics, 53 (1982) 5898-5903.

6. H. E. Cline, "Growth of Eutectic Alloy Thin Films", Materials Science and Engineering, 65 (1984) 93-100.

7. R. Elliott, Eutectic Solidification Processing, (Butterworth, London, 1983) 92-119.

8. W. J. Boettinger, "Surface Relief Cinemicrography of the Unsteady Solidification of the Lead-Tin-Cadmium Eutectic" (Ph.D. dissertation, The John Hopkins University, Baltimore, Maryland, 1972) 59-104.

9. T. Carlberg and H. Fredriksson, "On the Mechanism of Lamellar Spacing Adjustments in Eutectic Alloys", Journal of Crystal Growth, 42 (1977) 526-35.

10. V. Seetharaman and R. Trivedi, "Eutectic Growth: Selection of Interlamaller Spacings", Metallurgical Transactions (submitted).

11. J. T. Mason and M. A. Eshelman, "Model Transport Directional Solidification Apparatus", IS-4096, Ames Laboratory, Ames, Iowa, 1986.

12. P. G. Shewmon, Transformations in Metals, (McGraw Hill, New York, 1969) 156-208.

13. L. M. Hogan, R. W. Kraft and F. D. Lemkey, "Eutectic Grains", Advances in Materials Research, vol. 5, Ed H. Herman (John Wiley, New York, 1970) 83-216.

14. H. Dean and J. E. Gruzleski, "Observations on the Details of the Fault Line Movement in Lamellar Eutectics", Journal of Crystal Growth, 21 (1974) 51-57.

15. D. D. Double, "Imperfections in Lamellar Eutectic Crystals", Materials Science and Engineering, 11 (1973) 325-35.

# MICROSTRUCTURAL DEVELOPMENT AND MECHANICAL PROPERTIES

## OF Cu-Al IN-SITU FORMED COMPOSITES

M. L. Borg, J. J. Valencia and C. G. Levi

Materials Department
Department of Mechanical and Environmental Engineering
University of California
Santa Barbara, CA   93106

### Abstract

Highly deformed in-situ composites of a two-phase Cu-8.3%Al eutectic alloy have been produced by conventional casting (CC) and directional solidification (DS) followed by extensive mechanical working. As-cast microstructures
are dendritic/cellular and exhibit spacings on the order of 23 μm for conventional casting and 185 to 136 μm for directional solidification at 3 and
6 cm/hr, respectively. Cold working up to true strains of $\varepsilon = 5.3$ (99.5% CW)
was possible in the CC structure, leading to filament spacings as fine as
1 - 2 μm. Mechanical testing of samples deformed to $\varepsilon > 3$ revealed that the
strengths achieved in both CC and DS materials are higher than those observed
in any other composite systems explored to date, notably Cu-Nb and Cu-Fe, at
equivalent cold work strains. It is believed, however, that most of the
strengthening arises from work hardening of the matrix, with limited contribution of the second phase filamentary dispersion.

Solidification Processing of Eutectic Alloys
D.M. Stefanescu, G.J. Abbaschian and R.J. Bayuzick
The Metallurgical Society, 1988

## Introduction

In-situ formed filamentary composites are a class of materials produced by extensive deformation processing of two-phase alloys. Desirable microstructures consist of an extremely dense ($10^6$ - $10^{10}$ /cm$^2$) and uniform dispersion of very thin (5 - 100 nm) metallic filaments in a metallic matrix [1]. These composites are of interest primarily because of their enhanced work hardening behavior; achievable strengths are higher than those expected from a rule of mixtures calculation based on the strength of the individual constituents with the same amount of cold work strain [2]. Furthermore, the work hardening rates of these materials commonly increase with deformation up to true strains on the order of 10 (99.995% CW), whereas in most systems reinforced by non-deforming dispersoids the initially rapid work hardening rate decreases with increasing strain [3]. In addition to their high strengths, some of these composites may also exhibit attractive magnetic or superconducting properties [1].

The strengthening mechanisms at work in ultrafine filamentary composites are not completely understood, although several hypotheses have been put forward. In general, the strength of these alloys is thought to be associated with the high density of interfaces, but the role of those interfaces is still a subject of debate. It has been shown that the UTS exhibits a Hall-Petch type relationship with the interfilament spacing, suggesting that interfaces act primarily as barriers to dislocation motion [4,5]. On the other hand, it has been proposed that the most significant contribution to composite strength stems from the additional dislocations which are geometrically necessary to accommodate the strain incompatibility between the phases [3,6]. Regardless of the mechanism, it is generally agreed that interfilament spacings in the sub-micron range are desirable to optimize the interfacial area per unit volume and the resulting strength. It is also believed that the phases should be as mechanically disparate as possible, but sufficiently ductile to undergo the extensive deformation associated with composite fabrication [6].

By and large the composite systems most extensively investigated are based on a copper matrix with a bcc second phase, notably Cu-Nb [1,4,7-11], Cu-Fe [5,6,12,13], Cu-Cr and Cu-Mo [14]. Filaments as fine as 10 nm and interfiber spacings ranging from 100 nm in Cu-Nb [9] to a few micrometers in Cu-Fe [13] have been obtained by swaging and drawing arc-cast ingots containing 10 - 30 vol% of second phase. Concomitant room temperature strengths as high as 2230 MPa, or about G/20, have been obtained for Cu-18 vol% Nb deformed to a true strain of ∿11.5 (99.999% CW) [1]. This compares with maximum UTS values of ∿500 MPa for pure Cu drawn to similar cold work strains [10]. It is also significant that strengthening occurs in many cases below the minimum volume fraction of reinforcing phase required in conventional composites [1].

The present investigation deals with in-situ formed Cu-matrix composites produced by deformation processing of cast structures. The primary objective is to study the effects of solidification microstructure on the material behavior during forming and on the achievable strengths. It is anticipated that if the second phase could be aligned and refined prior to deformation, the interfilament spacings would be finer at any level of cold working strain, thus enhancing the resultant strengthening. While directional solidification appears to be the logical approach to control the size and orientation of the second phase, it has seldom been coupled with mechanical processing, and then only with limited success [2,15]. For example, DS of an Ag-28.1 wt% Cu eutectic alloy followed by wire-drawing up to 99.99% CW showed only a marginal increase in strengthening (<10%) when compared with conventionally cast material cold worked to the same strain [2].

88

Improvements in microstructural refinement and orientation are optimized in principle by plane-front coupled growth, where spacings may be an order of magnitude finer than those in dendritic/cellular microstructures [16]. However, most of the copper matrix alloys of interest are not suitable for this approach: Cu-Fe is a peritectic system, theoretically unsuitable for coupled growth [16], while Cu-Nb and Cu-Cr are simple eutectics, but the volume fraction of the second phase is minimal at the eutectic composition. Furthermore, all these systems have steep liquidi and the liquid-solid temperature range opens rapidly with increasing composition.

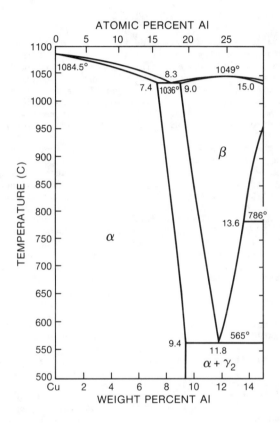

Figure 1 - Cu-rich end of the Cu-Al phase diagram from reference [17]. The eutectic composition is given in a more recent evaluation as 8 ± 0.3 %Al [18]. Note that all the alloys in the eutectic range (7.4 - 9.0 %Al) are single phase at room temperature.

A cursory review of the Cu-X binary phase diagrams reveals that the Cu-Al system selected for this study is the only Cu-matrix eutectic alloy involving a significant amount of a potentially ductile bcc second phase (i.e. the disordered intermetallic β). Figure 1 shows that this system has a reasonably low eutectic temperature and relatively short liquid-solid ranges on both sides of the eutectic composition. However, the (Cu) + β two-phase field is quite narrow making the microstructure evolution sensitive to small changes in Al content and/or solidification conditions. For example, a negative deviation of 0.5 %Al from the reported eutectic composition (8.3 %Al) would enhance the formability of the alloy as the volume fraction of second phase is cut in half, but it would also require a significant reduction in solidification velocity to maintain coupled growth. Further, the peculiar shape of the (Cu) + β phase field indicates that the 8.3%Al eutectic alloy should be single phase at temperatures below ∿1023 K (750 C). Thus, the second phase is prone to dissolve partially during cooling from the solidification temperature.

## Experimental

The Cu-Al alloy stock was prepared from electrolytic oxygen-free copper (99.999 %Cu) and high purity Al pellets (99.99 %Al). Melting was carried out by induction in a graphite-clay crucible coated with $ZrO_2$ mold wash. The melt was protected from oxidation by a $CaCO_3$ slag and was degassed for over 5 min. with argon injected through a graphite lance. The alloy was superheated 75 K above the eutectic temperature and cast into graphite molds having four 12 mm diameter, 150 mm long cylindrical cavities. The cast rods were cleaned and later used for directional solidification and/or deformation processing.

Given the characteristics of the phase diagram in Figure 1, it was deemed convenient to first explore the effects of Al content on volume fraction of β and growth morphology. To that effect, buttons ranging from 8.0 to 10.0 wt% Al were arc-melted in an argon atmosphere and examined metallographically. None of the microstructures exhibited regular coupled growth, as shown in Figure 2, but the 8.3 wt% Al button contained about 10 - 15 volume percent of β, which is on the lower end of the range used in most other composite systems (10-30 vol%) [1]. Since the second phase is a disordered intermetallic, conceivably with lower ductility than other reinforcements used so far, it was considered that larger amounts of second phase could hinder the formability of the material. Thus, it was decided to use the Cu-8.3wt%Al alloy for the initial stage of this investigation.

## Apparatus

The DS apparatus, shown in Figure 3, consists of a high gradient induction furnace heated by a cylindrical graphite susceptor and powered by a 20 kW, 400 kHz power supply. An alumina crucible (99.8% $Al_2O_3$, 13 mm I.D. and 305 mm long) containing the Cu-Al alloy is placed inside the susceptor. The whole furnace assembly is housed in a quartz chamber, where an argon flow is circulated from the top to protect both the susceptor and the liquid metal. A superheating of 100 K was typically achieved in the molten alloy away from the growing front; this temperature was held constant during the process. A temperature gradient of ∿50 K/cm (measured experimentally) was established on the liquid side of the interface by spraying cold water onto the crucible from a insulated stainless steel ring located at the bottom of the susceptor. The crucible is slowly withdrawn from the furnace using an Instron 1122 universal testing machine as a drive, which allowed for precise control of the solidification rate. Two solidification velocities, 3 cm/hr (8 µm/s) and 6 cm/hr (16 µm/s), were selected for the initial experiments based on previous experience with other alloy systems [e.g. 2].

## Deformation Processing

Prior to deformation processing the conventionally cast and directionally solidified rods were machined down to a diameter of ∿10 mm (400 mils) in order to remove any superficial oxides and casting defects. The rods were cold worked by swaging, limiting the reduction in area to less than 30% per step and constantly lubricating the dies with a light oil to keep the sample from overheating. The direction of swaging was reversed in each pass to prevent development of a spiral texture due to the rotating action of the dies [19]. 10 mm long samples for metallographic examination and hardness testing were taken at each swaging step beyond $\varepsilon = 1$. Material for about 5 tensile specimens was set aside at selected steps above cold work strains of 2.9.

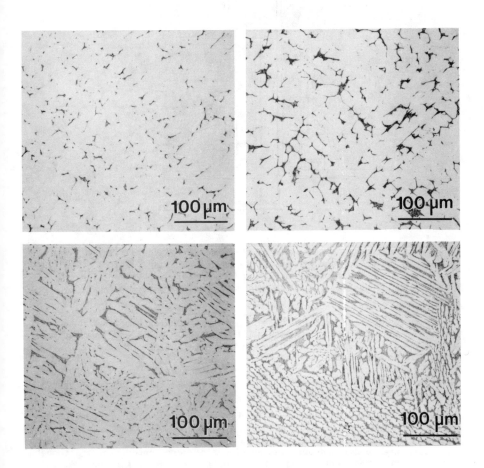

Figure 2 - Solidification microstructures of arc-melted Cu-Al buttons of nominal compositions (a) 8.0%Al, (b) 8.3%Al, (c) 8.6%Al and (d) 10%Al.

Characterization and Testing

Metallographic samples were mounted in conducting phenolic resin, mechanically polished and lightly etched with a $FeCl_3$/HCl-base solution to reveal the morphology of the second phase. Achieving a good-quality surface for optical examination was particularly difficult since the two-phase alloy does not lend itself to electropolishing or chemical polishing.

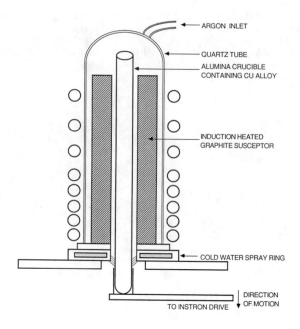

ARGON INLET

QUARTZ TUBE

ALUMINA CRUCIBLE
CONTAINING CU ALLOY

INDUCTION HEATED
GRAPHITE SUSCEPTOR

COLD WATER SPRAY RING

DIRECTION
OF MOTION

TO INSTRON DRIVE

Figure 3 - Schematic of the directional solidification furnace. The crucible is driven by an Instron
tensile testing machine and the graphite susceptor
is heated by a 20 kW, 400 kHz power supply.

Microindentation hardness testing was performed on samples mounted in
transverse orientation in bakelite and mechanically polished. Indentations
were made with a Vickers diamond pyramid indenter using a load of 200 grams.
Five readings were usually taken from each sample, one in the center and four
at half radius around the center. At the higher strains ($\epsilon > 2.9$), however,
the number of indentations was sometimes limited by incipient cracking and
the small cross sectional area of the specimens.

Tensile testing was performed on an Instron 1122 universal testing machine using standard grips and a crosshead speed of ~20 μm/s. In order to minimize stress concentrations at the edge of the grips, the ends of the composite wire were sandwiched between two pieces of soft copper sheet. Although
an extensometer was used on many of the tests, fracture often occurred outside the gauge length, preventing the accurate determination of yield
strength and elongation to failure. The smaller diameter specimens, especially those produced from DS material, often failed at preexisting cracks produced during swaging. In those cases the tests were repeated on the remaining
portion of the specimen until a reasonably clean fracture was produced.

## Microstructural Development

The solidification microstructures of the conventionally cast (CC) and
directionally solidified (DS) rods are shown in Figure 4. It is evident that
the interfacial conditions resulted in a dendritic or cellular morphology
rather than coupled growth, even though the alloy is at or very close to the
reported eutectic composition. In principle, one could argue that for growth

92

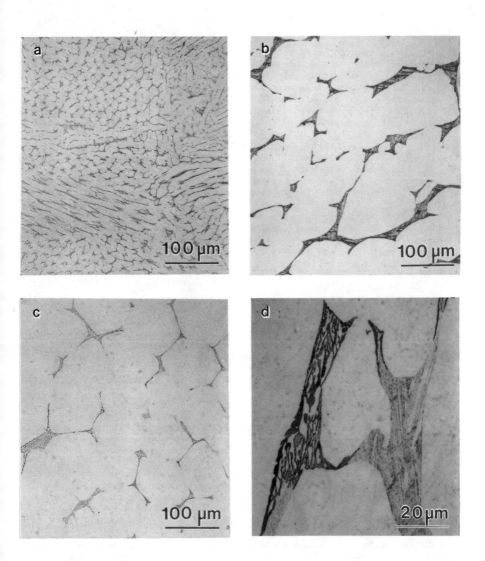

Figure 4 - Solidification microstructures of Cu-8.3%Al (a) conventionally cast, (b) directionally solidified at 6 cm/hr and (c) DS at 3 cm/hr; (d) shows a higher magnification view of a segregate region containing a mixture of (Cu) and β phases.

of (Cu) dendrites or cells to be favored at the eutectic composition, the coupled zone should be skewed towards the β field and the interfacial supercooling should fall out of the coupled zone in the phase diagram [20]. A skewed zone is normally associated with one of the phases growing in a faceted mode, which is not evident in any of the microstructures and would not be characteristic of either of·the disordered solid phases involved in this system. Furthermore, it is unlikely that any significant supercooling could be developed at the relatively low growth velocities used in directional solidification[1].

An alternate hypothesis is that the alloy used is really hypoeutectic, as suggested by the micrographs in Figure 2, where the 8.0, 8.3 and 8.6% alloys are seen to contain primary (Cu) with increasing amounts of $\beta$ segregate. EDXS analysis of the cast samples revealed the bulk composition of the alloy to be quite close to 8.3%Al, within the accepted range for the eutectic composition specified in a recent evaluation of the phase diagram [18]. On the other hand, the compositions of the primary and segregate phases were found to be ~7.7 and ~12 %Al, which are somewhat higher than the values indicated in Figure 1 but agree with the relative amounts of phases observed. Nevertheless, if the 8.3%Al alloy were indeed hypoeutectic, it should still be possible to produce coupled growth at much higher or much lower velocities than those used up to this point, provided that the coupled zone is reasonably symmetric. This approach will be further explored in future activities.

Figure 4 also shows that the segregate spacings in the directionally solidified rods are much larger than that in the conventionally cast one. As expected, the DS ingot grown at the faster velocity exhibits a finer spacing than that solidified more slowly, i.e. $\lambda$ = 136 $\mu$m and 185 $\mu$m for V = 6 and 3 cm/hr, respectively. It was rather surprising the product $\lambda^2$ V was reasonably constant,

$$\lambda^2 \, V \sim 3 \times 10^{-13} \, m^3/s \qquad (1)$$

since that type of relationship is predicted for coupled eutectics but not for cellular or dendritic growth. Although Equation (1) is not expected to hold over a wide range of velocities, it does suggest that the growth rate would have to be substantially increased (i.e. V ~ 200 cm/hr) in order to refine the DS structures to the level observed in the CC rods. Further increments, of course, could result in coupled growth with a more significant reduction in microstructural scale.

The larger segregate areas in the directionally solidified material, such as that in Figure 4(d), are sometimes found to be a mixture of phases rather than single phase $\beta$. This could be ascribed to coupled growth in the intercellular regions during solidification [20] or to solid state decomposition of the $\beta$ phase as it goes through the (Cu) + $\beta$ field during cooling. Although the issue has not been resolved, most of the $\beta$ regions are single phase and occasionally show evidence of a martensitic transformation, e.g. Figure 4(d), suggesting that the regions where phase separation is observed form during solidification. After deformation, these (Cu) + $\beta$ areas become bundles of second phase ribbons with some primary phase included.

Microstructure evolution during deformation processing is illustrated in Figure 5 for the CC and the DS rod solidified at 6 cm/hr. In the conventionally cast structure, the cold work progressively develops the alignment and refinement of the microstructure, reducing the interfiber spacing from 23 $\mu$m in the as-solidified structure to ~3 $\mu$m after a deformation strain of 3.3 (95% CW). Further cold working to a strain of 5.3 (99.5% CW) decreases the spacing to 1-2 $\mu$m. Although these spacings are comparable to those produced in some Cu-Fe composites [13], they are still coarser than those associated with the highest strengths obtained in Cu-Nb [1].

---

[1] Data for this system is not available, but relationships between velocity and supercooling established for the (Al)+CuAl$_2$ eutectic [21] suggest that $\Delta T$ < 1 K are typical of the velocities used in these experiments.

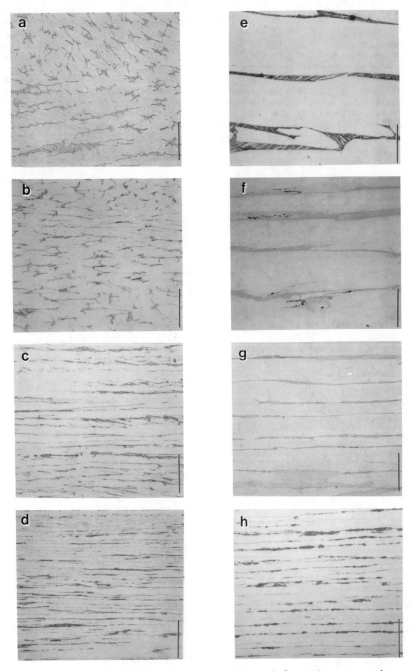

Figure 5 - Microstructure evolution during deformation processing in Cu-8.3%Al CC (left) and DS at 6 cm/hr (right). Cold work strains are 0 (a,e), 1.0 (b,f), 1.9 (c,g), and 3.3 (d,h). Marker bars are 50 μm for (a-d) and 100 μm for (e-h).

As expected from the solidification microstructures, much higher levels of deformation strain would be necessary in the DS rods in order to achieve the spacings observed in the CC microstructures. For example, the DS microstructure in Figure 5, cold worked to a strain of 3.3, exhibits a spacing comparable to that of the CC ingot in the as-cast condition. Further, the long residence time (several hours) in the liquid state associated with directional solidification promotes the formation of porosity defects, in spite of the efforts to control the environment in the DS furnace. In consequence, DS ingots could only be deformed to $\varepsilon$ = 3.3 before incipient cracking ensued, whereas the CC ingots were readily swaged down to $\varepsilon$ = 5.3. Thus, the limited benefits expected from pre-alignment of the second phase by DS were overcome by the increased scale and casting defect population of the microstructure, both of which are detrimental to the mechanical properties.

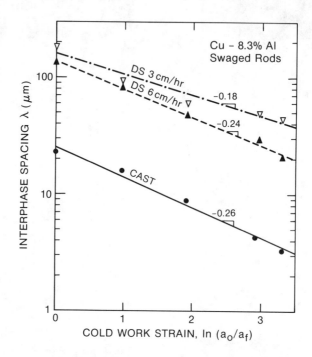

Figure 6 - Effect of deformation processing on interfiber spacing, $\lambda$, for conventionally cast and directionally solidified Cu-8.3%Al.

The effect of deformation processing on the interfiber spacing is depicted in Figure 6. A fairly linear trend is observed in all cases, with slopes close to -0.22 (-0.5/ln 10), or

$$\lambda = \lambda_o \exp (-0.5 \ \varepsilon) \tag{2}$$

This behavior is expected when the deformation is purely axisymmetric, i.e. if the scale of all the microstructural features decreases at the same rate as the cross sectional area of the rods. It is generally agreed that the bcc second phase in these composites undergoes plane strain deformation during cold working [22], as discussed below. However, the deformation is in general axially symmetric due to the higher number of slip systems active in the textured fcc Cu matrix [13,14]. Departures from axisymmetric behavior have been observed in Cu-Nb and ascribed to the increase in the density of second phase filaments on the cross section of the composite [10].

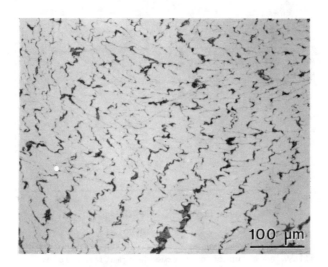

Figure 7 - Transverse view of the directionally solidified alloy (3 cm/hr) after a cold work strain of 3.3. Note the curled ribbon morphology of the second phase.

A transverse view of the microstructure after deformation reveals that the second phase assumes the curled ribbon morphology shown in Figure 7. The morphology arises from the <110> texture developed in bcc crystals during deformation [22]. When a <110> direction in a crystal becomes parallel to the rod/wire axis, there are only two <111> slip directions favorably oriented to accommodate the extension. Since these two directions lie on the same slip plane, further deformation becomes plane strain, producing the ribbon shape. The curling and kinking of the ribbons have been associated with the development of deformation bands within each crystal due to the different rotations of neighboring crystal portions to align a <110> with the wire axis [19].

## Mechanical Properties

Due to the limited availability of material for tensile testing, hardness measurements were used extensively in this work for a preliminary evaluation of the effects of cold working on mechanical properties. The measured hardness of the individual phases in the as-cast DS rod was 80 VHN for the matrix and 230 VHN for the $\beta$ phase. Reliable microhardness levels for each phase could not be determined in the CC rod due to the finer microstructure, but measurements of the overall hardness of a region containing both phases was found to be around 100 VHN, in reasonable agreement with the value predicted by the rule of mixtures.

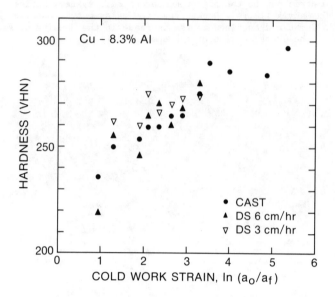

Figure 8 - Work hardening of Cu-8.3%Al CC and DS
alloys as a function of true deformation strain.

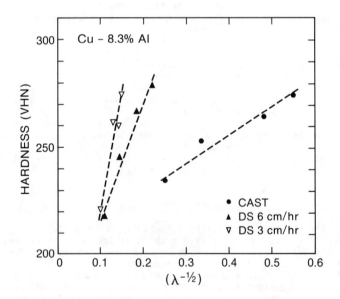

Figure 9 - Vickers hardness as a function of $\lambda^{-0.5}$
for CC and DS Cu-Al composites. The lowest values in
each line correspond to a cold work strain of 1.0,
whereas the highest values are for $\varepsilon = 3.3$.

The effect of the cold work strain on the VHN hardness for the conventionally cast and directionally solidified rods is shown in Figure 8. It was observed that initially the VHN increases rapidly with increasing deformation from ∿100 in the as-cast condition to about ∿255 at a strain of 1.3. The trend slows down as cold working proceeds, with the hardness increasing only up to ∿275 at a strains of 3.3, which was the maximum deformation achieved in the DS rods. The data fall within a relatively narrow band with no clear distinction between the hardness of the CC and DS alloys. Note also that the additional cold working of the CC material led to hardness values as high as 295 VHN.

Following the well established Hall-Petch relationship between UTS and interfiber spacing, we have plotted the VHN number against $1/\sqrt{\lambda}$ for the CC and DS materials in Figure 9. The trends are reasonably linear for the limited range of experimental values depicted. The lowest values correspond to a cold work strain of ∿1.0 (63% CW) and the highest to a strain of 3.3. As expected from the previous figure, the three composites have approximately the same hardness for equivalent strain levels in spite of the large differences in spacings. This suggests that the observed hardening may be due primarily to cold working of the matrix and that further refinement of the second phase may be necessary for it to have a significant effect.

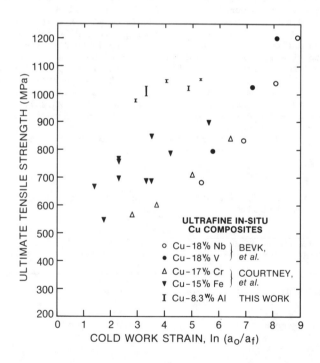

Figure 10 - Ultimate tensile strength as a function of cold work strain for the conventionally cast Cu-8.3%Al alloy compared with other systems reported in the literature [1,6].

The ultimate tensile strength of the conventionally cast material as a function of cold work strain is shown in Figure 10, along with results for other in-situ formed copper-matrix composites reported in the literature [1,6]. These preliminary results are encouraging since the UTS values are significantly higher than those of all the other systems at comparable levels of cold work. Nevertheless, the observed strengths are only about 1/2 of the maximum UTS achievable in the best Cu-Nb composites, which can be cold worked to much higher levels ($\varepsilon \sim 12$). It should also be noted that the present materials were deformed by swaging, which is likely to produce lower hardening rates than the drawing process commonly used for the other composites.

Table 1 compares the maximum strengths achieved in the DS Cu-Al composites with those of the CC rods and some relevant standard alloys. As expected from the microhardness data, the UTS values for both CC and DS materials at 95% cold work ($\varepsilon = 3.3$) are relatively close, in spite of the differences in spacing. Some insight on whether the matrix work hardening controls the strengthening of these composites may be gained by comparison with a single phase $\alpha$ bronze of similar composition (C61000). If strengthening is due to the matrix, one could argue that both materials should exhibit similar hardening, the differences in UTS arising from the more extensive deformation of the composites.

Comparison of the $\alpha$ bronze with pure copper in Table 1 indicates that the incremental strengthening of the matrix produced by the dissolved Al is 240 MPa in the annealed condition and 170 MPa after 37% CW ($\varepsilon = 0.46$). Consider now that pure Cu reaches a saturation UTS of $\sim 500$ MPa after extensive cold working [10]. The baseline strengthening produced by the matrix work hardening would then be 500 - 240 = 260 MPa. The large differences between the composite strengths and the saturation UTS of Cu, i.e. 420 to 560 MPa, would then be the incremental contribution of the solute. Aluminum additions above $\sim 4.5$ wt% have been shown to drastically change the character of the dislocation substructure in copper from cellular to coplanar arrays by reducing the stacking fault energy and hindering cross slip [23,24]. As a consequence, the Hall-Petch effect of the grain size is enhanced [23]. Thus, it is possible that the high composite strengths may result from the increased work hardening rate produced by the Al solute in the matrix, but the additional interfaces in the composite could also have a synergistic contribution to the strengthening.

Table 1 - Tensile strength of Cu-8.3wt%Al composites compared with those of standard Cu alloy rod/wire.

| Alloy | Processing | % C.W. | UTS (MPa) |
|---|---|---|---|
| Pure Cu (C10100) | Annealed | | 240 |
| | Drawn, H04 | 37 | 380 |
| Cu-8%Al (C61000) | Annealed | | 480 |
| | Drawn, H04 | 37 | 550 |
| Cu-8.3%Al | CC, Swaged | 99.5 | 1060 |
| | CC, Swaged | 95 | 980 |
| | DS 6cm/hr, Swaged | 95 | 920 |
| | DS 3cm/hr, Swaged | 95 | 950 |

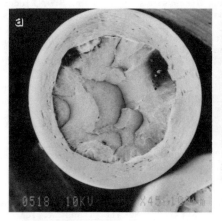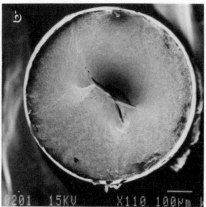

Figure 11 - Typical fracture surfaces of CC composites after cold
work strains of (a) 3.3 and (b) 5.3. The corresponding ductili-
ties are 46% and 10%, respectively.

Figure 11 depicts typical fracture surfaces of tensile test specimens at
two different cold work strains. In general, the fractures are microscopical-
ly dimpled and exhibit significant macroscopic ductility, but failure seems
to be commonly associated with existing cracks in the microstructure. These
could result from excessive deformation processing, as in Figure 11(b), but
may also arise from porosity and other casting defects. Smaller vertical
cracks are probably associated with the jagged appearance of the fracture
surface in Figure 11(a).

## Conclusions

The heavily cold-worked α-β 8.3% aluminum bronzes exhibit promising
strengths when compared with other fcc/bcc copper composites at similar deformation strains. UTS as high as 1060 MPa and hardness values on the order
of 295 VHN were achieved in these materials, but analysis of the mechanical
behavior suggests that the observed strengthening may be due primarily to
cold working of the matrix.

Whereas the conventionally cast material may be cold worked to 99.5%
($\varepsilon$ = 5.3), the directionally solidified alloys were only formable up to 95%
CW ($\varepsilon$ = 3.3), presumably due to an increased population of casting defects.
Spacings as fine as 1-2 μm were achieved in the CC material, but directional
solidification at growth rates up to 6 cm/hr resulted in significantly coarser microstructures in the as cast and deformed conditions.

Future efforts in this program are first aimed at clarifying the contribution of the second phase to composite strengthening by processing a single
phase alloy of the same composition at comparable cold work strains and
studying its mechanical behavior. A second goal is to pursue the development
of a finer microstructure by directional solidification, perhaps with minor
adjustments in composition. Finally, an alternate route based on rapid solidification, powder metallurgy and deformation processing will be explored.

## Acknowledgements

The authors are grateful to Prof. T.H. Courtney, Dr. J.C. Malzahn-Kampe and Prof. R. Mehrabian for enlightening discussions, and to Dr. S.D. Ridder and Mr. F. Biancaniello of the National Bureau of Standards for their assistance with the deformation processing experiments. This work was sponsored by the Defense Advanced Research Projects Agency under Grant N00014-86-K-0178, monitored by the Office of Naval Research. Program Director at DARPA is Dr. P. Parrish and contract monitor at ONR is Dr. S. Fishman.

## References

1.  J. Bevk, "Ultrafine Filamentary Composites", Ann. Rev. Mater. Sci., 13 (1983), 319-38.

2.  G. Frommeyer and G. Wassermann, "Microstructure and Anomalous Mechanical Properties of In-Situ Produced Silver-Copper Composite Wires", Acta Metall., 23(11)(1975), 1353-1360.

3.  P.D. Funkenbusch, J.K. Lee and T.H. Courtney, "Ductile Two-Phase Alloys: Prediction of Strengthening at High Strains", Metall. Trans. A, 18A (1987), 1249-1256.

4.  J. Bevk, J.P. Harbison and J.L. Bell, "Anomalous Increase in Strength of In-Situ Formed Cu-Nb Multifilamentary Composites", J. Appl. Phys., 49 (12)(1978), 6031-38.

5.  P.D. Funkenbusch and T.H. Courtney, "Microstructural Strengthening in Cold Worked In-Situ Cu-14.8 v/o Fe Composites", Scripta Metall., 15 (1981), 1349-1354.

6.  P.D. Funkenbusch and T.H. Courtney, "On the Strength of Heavily Cold Worked In-Situ Composites", Acta Metall., 33(5)(1985), 913-922.

7.  H.E. Cline, B.P. Strauss, R.M. Rose and J. Wulff, "Superconductivity of a Composite of Fine Niobium Wires in Copper", J. Appl. Phys., 37(1) (1966), 5-8.

8.  J. Bevk and K.R. Karasek, "High Temperature Strength and Fracture Mode of In-Situ Formed Cu-Nb Multifilamentary Composites", in New Developments and Applications in Composites, ed. D. Kuhlmann-Wilddorf and W.C. Harrigan, (Warrendale, PA: The Metallurgical Society, 1979), 101-113.

9.  D.E. Cohen and J. Bevk, "Enhancement of the Young's Modulus in the Ultrafine Cu-Nb Filamentary Composites", Appl. Phys. Lett., 39(8) (1981), 595-597.

10. W.A. Spitzig, A.R. Pelton and F.C. Laabs, "Characterization of the Strength and Microstructure of Heavily Cold Worked Cu-Nb Composites", Acta Metall., 35(10) (1987), 2427-2442.

11. P.J. Lee and D.C. Larbalestier, "Development of Nanometer Scale Structures in Composites of Nb-Ti and Their Effect on the Superconducting Critical Current Density", Acta Metall., 35(10)(1987), 2523-2536.

12. F.P. Levi, "Permanent Magnets Obtained by Drawing Compacts of Parallel Iron Wires", J. Appl. Phys., 31(8)(1960), 1469-1471.

13. J.C. Malzahn-Kampe and T.H. Courtney, "Elevated Temperature Microstructural Stability of Heavily Cold-Worked In-Situ Composites", Scripta Metall., **20** (1986), 285-289.

14. P.D. Funkenbusch, T.H. Courtney and D.G. Kubisch, "Fabricability and Microstructural Development in Cold Worked Metal Matrix Composites", Scripta Metall., **18**(1984), 1099-1104.

15. D.G. Kubisch and T.H. Courtney, "The Processing and Properties of Heavily Cold Worked Directionally Solidified Ni-W Eutectic Alloys", submitted for publication.

16. W. Kurz and R.J. Fisher, Fundamentals of Solidification, (Aedermannsdorf, Switzerland: Trans Tech Publications, 1984).

17. T. Lyman et al., eds., Metals Handbook, vol. 8, (Metals Park, OH: American Society for Metals, 1973), 259.

18. J.L. Murray, "The Al-Cu Phase Diagram", Int. Met. Rev., **30**(5), 1985, also in Binary Alloy Phase Diagrams, ed. T.B. Massalski, (Metals Park, OH: ASM International, 1986), 103.

19. J.C. Malzahn Kampe: Ph.D. Thesis, Michigan Technological University, 1987.

20. W. Kurz and D.J. Fisher, "Dendrite Growth in Eutectic Alloys: The Coupled Zone", Int. Met. Rev., 1979, Nos. 5 and 6: 177-203.

21. S.M.D. Borland and R. Elliot, "Growth Temperatures in Al-CuAl$_2$ and Sn-Cd Eutectic Alloys", Metall. Trans. A, **9A** (1978), 1063-1067.

22. W.F. Hosford, Jr., "Microstructural Changes During Deformation of [011] Fiber-Textured Metals", Trans. AIME, **230** (1964), 12-15.

23. T.L. Johnston and C.E. Feltner, "Grain Size Effects in the Strain Hardening of Polycrystals", Metall. Trans., **1** (1970), 1161-1167.

24. R.W.K. Honeycombe, The Plastic Deformation of Metals, (London, UK: Edward Arnold Publishers, Ltd., 1968).

STRUCTURES IN DIRECTIONALLY SOLIDIFIED

ALUMINUM FOUNDRY ALLOY A356

D. A. Granger* and E. Ting**

*Alcoa Laboratories/**Grumman Research Center
Alcoa Center,               Bethpage,
   PA   15069               NY   11714

## ABSTRACT

In preparation for conducting directional solidification experiments of a commercial aluminum casting alloy under conditions of microgravity, a series of ground-based tests have been conducted employing controlled temperature gradient in the liquid and interface velocity. The alloy selected was A356 which was prepared in the unmodified, strontium modified and antimony refined condition. Growth conditions were selected to give a range of morphologies as well as being achievable with the anticipated capabilities of the AADSF (advanced automated directional solidification furnace).

Structures developed in these experiments were examined using optical and scanning electron microscopy. Metallurgical factors such as dendritic cell size and eutectic particle refinement are interpreted in terms of the growth conditions and effect of the chemical modifiers.

Solidification Processing of Eutectic Alloys
D.M. Stefanescu, G.J. Abbaschian and R.J. Bayuzick
The Metallurgical Society, 1988

## Introduction and Background

The experiments described in this report were selected as part of a joint endeavor between Alcoa Laboratories and Grumman R&D to explore the impact of microgravity growth conditions on the structure (and properties) of directionally solidified Al-Si-Mg alloy samples. Solidification conditions were chosen that would be compatible with those obtainable in the AADSF (Advanced Automatic Directional Solidification Furnace) in order to allow a direct comparison between cast samples prepared on earth with those produced under conditions of microgravity (on board a Shuttle or Space Station). Alloy selection was based on several criteria including: (1) closely approximating a commercial casting alloy, (2) simple enough to allow comparison with results of similar experiments on binary alloys, (3) markedly different structures are generated, yet not fully understood, by trace element additions and (4) there is an interacting influence of growth conditions and chemical modification.

Interpretation of results, including validation of the growth conditions, could be carried out by making use of published data. For example, it is well established that the dendrite cell size is directly related to the solidification rate (1,2). Thus, the conditions of controlled growth rate (R) and temperature gradient (G) allow the relationship between dendrite cell size and solidification rate to be used to verify the growth conditions. Furthermore, the eutectic particle spacing in the selected alloy was expected to respond to growth conditions in a similar fashion to the extensively studied Al-Si binary alloy (3-5).

Producing samples under gravity and microgravity conditions may be anticipated to shed light on the underlying mechanism of structural refinement conferred by trace additions of antimony and structure modification brought about by strontium.

## Experimental Procedures

### Preparation of Charge

The experiments were conducted using an Al-6.5% Si-0.45% Mg alloy in the modified, unmodified and refined conditions. Compositions of the specimens used are given in Table I. Samples for the controlled solidification

Table I.  Chemical Composition of A356 Type Alloys

| Alloy* | Modification/Refinement | Si | Mg | Sr | Sb |
|--------|------------------------|-----|------|-------|------|
| A | None | 6.4 | 0.44 | 0.000 | 0.00 |
| B | Sr | 6.7 | 0.45 | 0.030 | 0.00 |
| C | Sb | 6.6 | 0.46 | 0.000 | 0.06 |

*Composition given in weight percent.

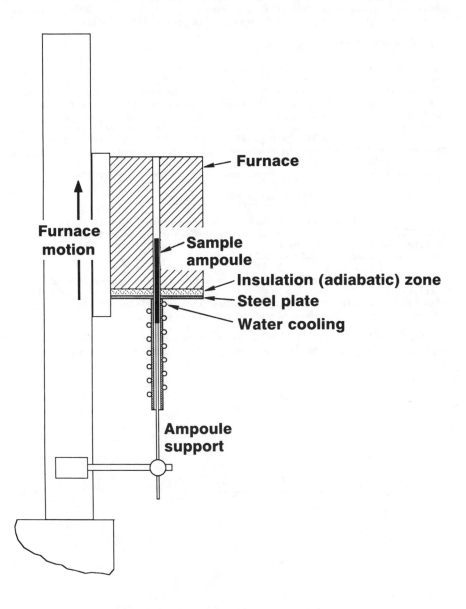

Figure 1. Directional Solidification Furnace

experiments were prepared from a melt made up in a zirconite coated plumbago crucible heated in an electric resistance furnace. After melting the charge the temperature was stabilized at 1300°F (704°C) and the melt fluxed with an Ar-10% $Cl_2$ until a hydrogen content of less than 0.12 ml/100 g had been obtained; samples were then taken for chemical analysis and 1/2 in. x 12 in. long (12 x 300 mm) castings were prepared using a split graphite mold.

## Directional Solidification Experiments

The gradient furnace in which the directional solidification experiments were conducted is illustrated in Figure 1. The charge, which was contained in a mullite mold, comprised a cylindrical specimen about 75 mm long x 4 mm dia. After remelting and solidifying under the controlled solidification conditions, i.e., fixed temperature gradient in the melt and constant interface velocity, the specimens were sectioned and prepared for metallographic examination both parallel and normal to the growth direction. Measurements were made of the dendrite cell size and interparticle spacing on all the specimens. In addition, the samples were examined by scanning electron microscopy (SEM) to better reveal the growth morphologies.

## Results and Discussion

All the specimens were prepared in the directional solidification furnace with a temperature gradient of 400°C/in. (16°C/mm) and interface velocities ranging from 0.30 to 3.0 in./h (0.002 to 0.02 mm/s). Thus solidification rates used varied from 0.03 to 0.33°C/s.

An A356 type alloy composition was used for all the experiments with modification (refinement) as listed in Table I. The composition was only adjusted by addition of elements to induce either modification (Sr) or refinement (Sb) of the eutectic silicon particles.

Figure 2 illustrates the manner in which the specimens were sectioned for subsequent examination by optical and scanning electron microscopy (SEM).

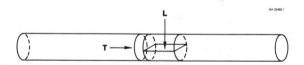

Figure 2. Schematic showing how transverse (T) and longitudinal (L) sections were cut from casting.

108

Structures in the unmodified alloy grown at the slowest 0.30 in./h. (0.002 mm/s) and the fastest rate 3.0 in./h. (0.02 mm/s) are illustrated in Figure 3(a-d). At the slowest rate a gas hole formed that propagated the full length of the specimen. This was the only specimen to exhibit such a defect and was not, in this sample, expected to perturb the growth structure in terms of cell and eutectic particle (or flake) spacing. Figure 4(a-d) is a similar set of photomicrographs showing the influence of a strontium addition to induce chemical modification and Figure 5(a-d) shows the effect of antimony which refines the spacing of the eutectic silicon lamellae. All the specimens were grown under identical solidification conditions.

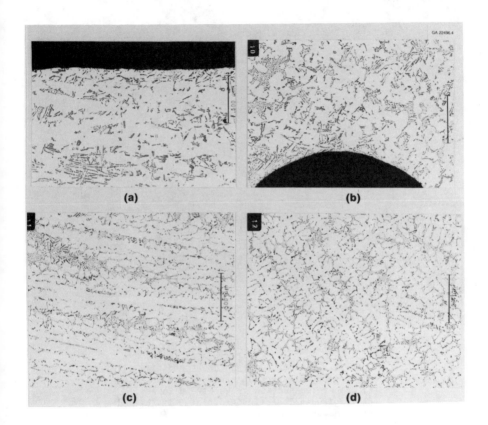

Figure 3. Photomicrographs of unmodified alloy.
(a) longitudinal section grown at 0.3 in./h.,
(b) transverse section grown at 0.3 in./h.,
(c) longitudinal section grown at 3 in./h. and
(d) transverse section grown at 3 in./h.

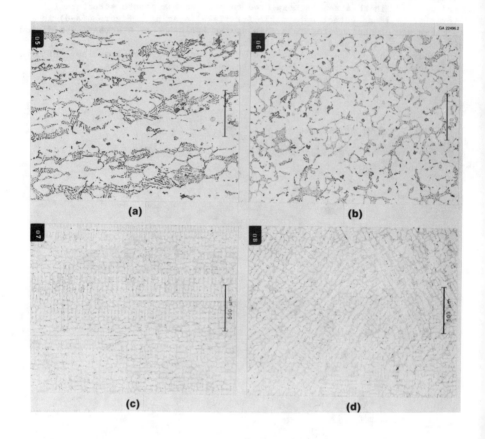

Figure 4. Photomicrographs of Sr modified alloy.
(a) longitudinal section grown at 0.3 in./h.,
(b) transverse section grown at 0.3 in./h.,
(c) longitudinal section grown at 3 in./h. and
(d) transverse section grown at 3 in./h.

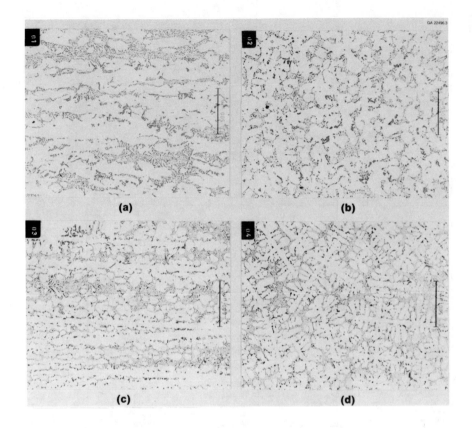

Figure 5.  Photomicrographs of Sb refined alloy.
(a) longitudinal section grown at 0.3 in./h.,
(b) transverse section grown at 0.3 in./h.,
(c) longitudinal section grown at 3 in./h. and
(d) transverse section grown at 3 in./h.

    The influence of cooling rate and chemical addition on the eutectic
structure is illustrated at higher magnification in Figure 6(a-f).  The
difference in the effect of Sr and Sb is particularly evident in the
photomicrograph illustrating the specimens grown at the highest interface
velocity.

111

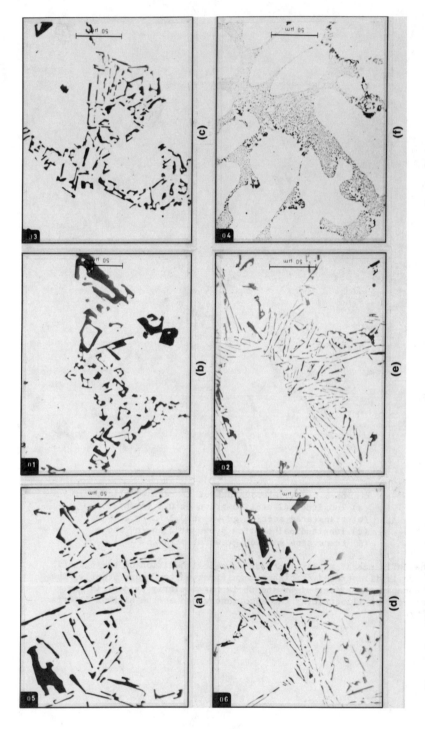

Figure 6. Photomicrographs showing eutectic structure in (a) unmodified, (b) Sb refined, (c) Sr modified specimens grown at 0.3 in./h., (d) unmodified, (e) Sb refined and (f) Sr modified specimens grown at 3 in./h.

In order to better illustrate the effect of (1) interface velocity (or solidification rate) and (2) chemical additives, SEMs were prepared from the specimens after they had been etched with a bromine-ethanol solution to remove the aluminum matrix. The first three illustrations, Figures 7-9, show the structures with no modification, Sr modification and Sb refinement prepared under conditions of slowest freezing (0.03°C/s). And the second set, Figures 10-12, are taken from the specimens grown at the highest solidification rate. These SEMs demonstrate the powerful influence solidification rate (or interface velocity) has on enhancing the effect of the chemical additives. Figure 11 shows most dramatically the influence of Sr in producing a continuous fibrous eutectic structure while Figure 12 shows how Sb generates a plate-like (or lamellar) structure.

Quantitative Metallography

Cell size measurements were made using a linear intercept method described by Spear and Gardner (1) on both the transverse and longitudinal specimens. The results of these measurements are listed in Table II.

Table II:  Calculated and Measured Cell Size
Measurements

| R (in./h.) | G (°C/in.)** | Solidification Rate(°C/S) | Cell Size (μm) Calculated* | Measured |
|---|---|---|---|---|
| 0.3 | 400 | 0.03 | 111 | 109 ± 5 |
| 0.7 | 400 | 0.08 | 81 | 82 ± 5 |
| 1.7 | 400 | 0.19 | 61 | 57 ± 5 |
| 3.0 | 400 | 0.33 | 50 | 56 ± 5 |

* Using d (cell size) = 35 $\theta^{-0.33}$
  where $\theta$ is the solidification rate.

**157°C/cm.

The measured results were obtained by averaging values from the alloy with and without elements added for modification and refinement since the effect of these elements on the cell size was within the experimental error. The calculated results were obtained using the expression:

$$d = 35\ \theta^{-0.33}$$

where d = cell size (μm) and $\theta$ = solidification rate (°C/s).

Figure 7. SEM illustrating Si phase in unmodified alloy grown at 0.3 in./h.

Figure 8. SEM illustrating Si phase in Sr modified alloy grown at 0.3 in./h.

Figure 9. SEM illustrating Si phase in
Sb refined alloy grown at 0.3 in./h.

Figure 10. SEM illustrating Si phase
in unmodified alloy grown at 3 in./h.

Figure 11.  SEM illustrating Si phase
in Sr modified alloy grown at 3 in./h.

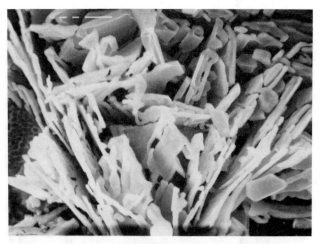

Figure 12.  SEM illustrating Si phase
in Sb refined alloy grown at 3 in./h.

116

The good agreement between the measured and calculated cell sizes gives confidence that the solidification conditions were accurately determined and that the specimens grew with a flat interface in the adiabatic zone of the furnace (see Figure 1).

Interparticle spacings were measured in all the samples and are listed in Table III. For the purpose of comparing the data generated in this work and that published in the literature, see for instance (5,6), the interparticle spacing was plotted as a function of interface velocity for the unmodified and the Sr modified alloy. The relationship between the interparticle spacing and interface velocity is in good agreement with the previously published work of Atasoy (6) but slope of the curve in Figure 13 is much steeper for the modified alloy. There is no explanation for this result at the present time, but it does suggest that there is an influence of magnesium in this hypo-eutectic alloy.

Table III. Interparticle Spacing as a Function of
Solidification Rate and Modification in
Al-6.5 Si-0.45 Mg Alloy

| Rate (°C/S) | Addition | Interparticle Spacing (μm) |
|---|---|---|
| 0.03 | None | 16.0 |
| 0.03 | Sb | 11.7 |
| 0.03 | Sr | 9.3 |
| 0.08 | None | 12.0 |
| 0.08 | Sb | 11.0 |
| 0.08 | Sr | 4.0 |
| 0.19 | None | 8.0 |
| 0.19 | Sb | 10.0 |
| 0.19 | Sr | 2.0 |
| 0.33 | None | 6.9 |
| 0.33 | Sb | 6.5 |
| 0.33 | Sr | 1.2 |

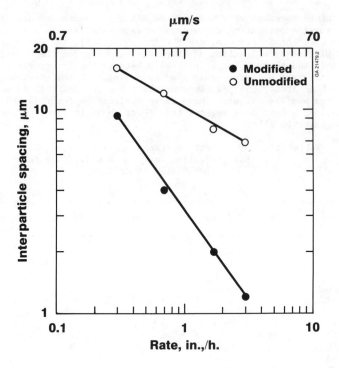

Figure 13. Variation in Log λ with
Log R for G = 400°C/in. (157°C/cm).

## Conclusions

1. The cell size decreases with increasing solidification rate in accordance with the relationship, $d = k\theta^{-0.33}$.

2. Cell size is not affected by the introduction of either a chemical modifier or refiner into the alloy.

3. Both Sr and Sb were retained on remelting and resolidifying. Their effects on the eutectic structure were as anticipated.

4. Strontium was more effective in reducing interparticle spacing than antimony under the range of growth conditions investigated.

5. Interparticle spacing is in good agreement with measurements made on unmodified Al-Si eutectic alloy, but not with the modified alloy.

## References

1. R. E. Spear and G. R. Gardner, Modern Castings, "Dendrite Cell Size," 37, (1960) 32-44.

2. J. A. Horwath and L. F. Mondolfo, Acta Met., "Dendritic Growth," 10, (1962) 1037-1042.

3. R. Elliott, Eutectic Solidification Processing, Butterworths, London (1983), 157.

4. M. D. Hanna, Shu-Zu Lu and A. Hellawell, "Modification in the Aluminum Silicon System," Met. Trans. A., 15A, (1984) 459-469.

5. L. M. Hogan and H. Song, "Interparticle Spacings and Undercoolings in Al-Si Eutectic Microstructures," Met. Trans. A., 18A, (1987) 707-713.

6. O. A. Atasoy, "Effects of Unidirectional Solidification Rate and Composition on Interparticle Spacing in Al-Si Eutectic Alloys," Aluminium, 60, (1984) 275-278.

# DIRECTIONAL SOLIDIFICATION OF Si- AND GaAs-BASED EUTECTICS FOR ELECTRONIC APPLICATIONS

B.M. Ditchek, T.R. Middleton, K.M. Beatty, and J.A. Kafalas

GTE Laboratories Incorporated
40 Sylvan Road
Waltham, MA 02254

## Abstract

The unidirectional solidification of most eutectic melts has been performed in a vertical Bridgman mode. In this geometry, the temperature gradient is stabilizing, melt convection is laminar, and the eutectic microstructure grows almost unaffected by melt flow. Si- and GaAs-based eutectics have also been predominantly grown by Bridgman-type techniques. This was the basic growth approach used by L. Levinson (1) in the Si-CrSi$_2$ system, Helbren and Hiscocks (2) in several other Si-silicide systems, and by Reis and Renner (3) in GaAs-metal arsenide eutectic systems. For all of these semiconductor-metal eutectic (SME) systems, the semiconductor matrix was not seeded and therefore grew into a polycrystalline, twinned material. Of course, the growth of pure, single-phase Si and GaAs semiconductor substrates almost always uses the Czochralski growth technique because of the ease of single-crystal seeding and the reduction of stresses and impurities. For these semiconductors the benefits of Czochralski growth outweigh the disadvantages associated with the turbulent melt convection induced by the destabilizing temperature gradients. It is becoming apparent that crystallinity, purity, and stress relief must also be considered in the growth of eutectic composites if such materials are to be useful in electronic applications.

Recently, we successfully applied the Czochralski crystal growth process to the growth of Si-TaSi$_2$ rod-like eutectics and obtained composites with a single-crystal Si matrix of electronic quality (4,5). These materials were used to fabricate diodes (4), photodiodes (6), and transistors (7) that utilized the *in-situ* Schottky junctions grown in the composites.

A determination of the range of possible microstructural and electrical properties of this new class of materials is important for its continued development. Microstructural factors that affect the application of these materials in devices include rod shape, density, alignment, and volume fraction. Other considerations are the Schottky barrier height between the semiconductor and metal phases and the control of the level of electrically active impurities and native defects. Further realization of a wide range of eutectic devices will depend on the development of other eutectic systems that enable variation of some or all of these parameters.

In this paper, we examine the Czochralski growth of the Si-TaSi$_2$ eutectic as well as the growth of other Si-silicide and GaAs-metal arsenides. Particular emphasis is placed on the relevant microstructural features of these eutectics as they are affected by the Czochralski growth technique.

Solidification Processing of Eutectic Alloys
D.M. Stefanescu, G.J. Abbaschian and R.J. Bayuzick
The Metallurgical Society, 1988

## Experimental Methods

The Si-based SME composites were grown in an RF heated crystal puller. The charge is placed in a quartz crucible surrounded by a graphite susceptor. A Si(111) seed was used in all cases. During growth in 1 atm of flowing Ar, the crucible and seed crystal were rotated in opposite directions at 6 rpm. Growth rates between 2 and 20 cm/h were employed. Eutectics between the disilicides of W, Ta, Nb, Zr, Cr, and Co with Si were examined. The resistivity of the Si charge was 10 $\Omega$-cm or higher. The purity of the metals specified by the vendors is given in Table I. Temperature gradients in the unseeded melt were about 100 °C/cm as measured by a W-Re thermocouple.

Table I. Purity of Metallic Elements

| Element | Purity |
|---------|--------|
| W | 99.999 |
| Mo | 99.99 |
| Ta | 99.996 |
| Nb | 99.985 |
| Zr | 99.95 |
| Cr | 99.999 |
| Co | 99.998 |

A Malvern MRS-6 high-pressure puller was used to grow the GaAs-based eutectics. This is a commercial system designed for LEC growth of GaAs boules about 5 cm in diameter. The Ga and As, both 99.99999 % pure, were placed in a pyrolytic BN crucible along with the metal of choice (either Mo or Cr) and capped with a plate of $B_2O_3$. The chamber was evacuated with a roughing pump and then pressurized to 800 psi with Ar. This pressure is maintained until the molten eutectic composition is formed and then reduced to 300 psi for growth. Reduction of the pressure immediately following reaction of the Ga and As as is normally done in GaAs crystal growth results in loss of the free As and a deviation in composition away from the eutectic. A (111) GaAs seed was used. Both seed crystal and crucible were rotated in opposite directions at a rate of 10 rpm. Boules were pulled at 6 cm/h.

## Si-Based Eutectics

### Survey of Microstructures

Microstructures of the Si-based eutectics are shown in Figure 1: All the scanning electron microscope (SEM) images of wafers cut transverse to the growth direction were obtained from samples etched in a NaOH/NaOCl solution to remove the Si and expose the disilicide phases. The composite examined in this survey all had a polycrystalline Si matrix even though a Si seed was used. Conditions for obtaining single-crystal Si matrices are discussed at the end of this section. It should be noted that the microstructure of the eutectics did not depend on orientation of the Si. In the Si-TaSi$_2$ system, there was no significant microstructural difference between composites with a single-crystalline and a polycrystalline Si matrix.

The microstructures presented in Figure 1 are in order of increasing volume fraction of the disilicide phases. The W, Ta, and Nb disilicides lie in the low volume fraction end and display similar microstructures. The structures are predominantly rod-like with substantial faceting of the disilicide phases. That is, the rods are not cylindrical, but rather have

122

cross sections that are either triangular, hexagonal, or rectangular (plate-like) in shape. The rods are well aligned, but some divergence does exist. In certain cases, it has been noted that extended or plate-like rods have resulted from the intersection of two closely spaced, misaligned rods some distance below the surface.

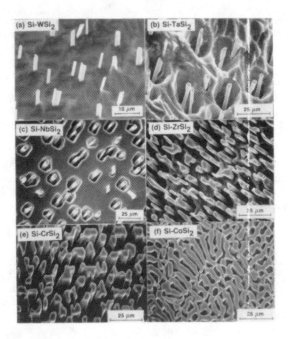

Figure 1.  Scanning electron micrographs of Si-based eutectics. The Si phase has been etched back to reveal the silicide phase. All composites were grown at 2 cm/h except for the Si-WSi$_2$ eutectic, which was pulled at 10 cm/h.

The microstructures are quite similar to the Bridgman-grown Si-TaSi$_2$ and Si-NbSi$_2$ structures observed by Helbren and Hiscocks (2). For the Si-NbSi$_2$ system, Helbren and Hiscocks reported rods of triangular cross section. Similar structures have occasionally been observed in the Czochralski-grown material and, similar to their findings, are generally larger than the cylindrically shaped silicide phase. Micrographs taken along the longitudinal axis of their Bridgman-grown boules show some misalignment of rods. A maximum 6° divergence of the rods from the growth axis that we have observed for the Si-TaSi$_2$ system (5) appears to be in reasonable agreement with the micrographs shown by Helbren and Hiscocks (2).

The similarity of the Si-TaSi$_2$ and the Si-NbSi$_2$ microstructures grown by Bridgman and Czochralski indicate that turbulence in the Czochralski melt is not adversely affecting their microstructures. The only other previous comparison of a Czochralski and Bridgman eutectic was in the Ge-TiGe$_2$ system (8). The Bridgman-grown eutectic exhibited TiGe$_2$ rods with a uniform rod diameter. The Czochralski-grown composites, on the other hand, displayed TiGe$_2$ rods with oscillations in the rod diameter that correlated with seed and crucible rotation rates. This distinction between the shape of the metallic rods in the Czochralski versus Bridgman material was not evident in these Si-based rod-like eutectic systems.

The Si-ZrSi$_2$, Si-CrSi$_2$, and Si-CoSi$_2$ eutectics have 19, 29, and 55 volume % disilicide phases. The higher volume fraction eutectics show extensive interconnectivity of the silicide phases. The basic unit shape of the silicide phases is very similar to that for the low volume fraction silicide phases, but the higher volume fraction makes the occurrence of intersection more common and causes the phases to appear extended. At the high end of the volume fraction range, typified by the Si-CoSi$_2$ system, the phase distribution is more lamellar than rod-like.

Levinson has previously grown the Si-CrSi$_2$ eutectic by the Bridgman technique (1). The microstructure obtained was very similar to the Czochralski-grown microstructure; both grown techniques resulted in the intersection of rods to form extended networks of the silicide phase.

### Growth Rate Effects

The spacing of the phases of regular eutectics, $\lambda$, depends on the growth rate according to

$$\lambda^2 v = A, \tag{1}$$

where A, a constant for a particular system, is primarily determined by the interfacial energy of the two phases and the diffusion coefficient in the melt (9). A similar expression may be written for the dependence of d, the rod diameter or characteristic width of a lamallae, on growth rate,

$$d^2 v = B, \tag{2}$$

where B is a constant, since d is proportional to $\lambda$. Values of $\lambda^2 v$, $d^2 v$, f, the volume fraction of the silicide phase, and a calculated value of d at 10 cm/h for each of the Si-based eutectic systems examined in this study are shown in Table II. Each system was grown at a minimum of two growth rates. The Si-TaSi$_2$ (10) and the Si-CoSi$_2$ (11) systems have previously been shown to observe the $\lambda^2 v$ rule over a wide range of growth rates.

Table II. Microstructural Parameters for Si-Silicide Eutectics

| Eutectic | f(%) | $\lambda^2 v$ ($10^{-8}$cm$^3$/h) | $d^2 v$ ($10^{-8}$cm$^3$/h) | d ($10^{-4}$cm) at v = 10 cm/h |
|---|---|---|---|---|
| Si-WSi$_2$ | 0.9 | 1150 | 20 | 1.4 |
| Si-TaSi$_2$ | 2.2 | 1250 | 13 | 1.1 |
| Si-NbSi$_2$ | 2.8 | 1050 | 30 | 1.7 |
| Si-ZrSi$_2$ | 19 | 75 | 17 | 1.3 |
| Si-CrSi$_2$ | 29 | 90 | 30 | 1.7 |
| Si-CoSi$_2$ | 55 | 90 | 18 | 1.3 |

An examination of the values of $\lambda^2 v$ for each system shows that the constant decreases as the volume fraction of the silicide phase increases. The value of $d^2 v$, which is approximately $\lambda^2 v$ multiplied by the silicide volume fraction, does not vary significantly within the Si-silicide eutectic systems. Values of d calculated for a 10 cm/h growth rate show that for each system, d = 1.4 ± 0.3 $\mu$m. This similarity of the eutectic spacings, along with the similarity of the eutectic microstructures noted in the previous section, suggests that factors like the interfacial energy and the diffusion coefficient are quite similar for all the Si-silicide eutectics examined.

124

The shape of the silicide phases also depends on growth rate. We observed that the frequency of triangular and other irregularly shaped rods decreased as the growth rate increased over the 2 cm/h to 20 cm/h range. The trend toward increased regularity with growth rate was observed for all the eutectics examined. Typical micrographs taken for the Si-NbSi$_2$ system following growth at 2 cm/h and 20 cm/h are given in Figure 2. For this system an image analysis was performed based on over 10 picture frames similar to those in Figure 2. The analysis yielded an average rod density of $2.2 \times 10^5$ rods/cm$^2$ at 2 cm/h and $2.6 \times 10^6$ rods/cm$^2$ at 20 cm/h. The increase in rod density is commensurate with Equation 1. Comparison of the individual micrographs demonstrates that the microstructural uniformity of the eutectic increases as the average interrod spacing decreases. While arrow-shaped rods and thin plates are found commonly in composites grown at 2 cm/h, these features are rare in composites pulled at 20 cm/h.

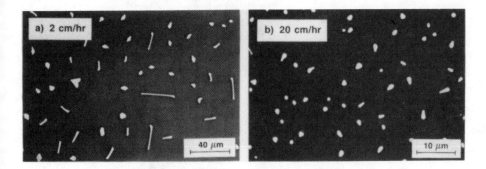

Figure 2. SEM micrographs of the polished surface of transverse sections of a Si-NbSi$_2$ eutectic pulled at (a) 2 cm/h and (b) 20 cm/h. The faster pull rate yields a more regular microstructure.

### Single-Crystal Growth

Growth of a single-crystal matrix eutectic depends primarily on using the exact eutectic composition and minimizing certain impurities. It is highly unusual to start with the exact eutectic composition due to the loss of Si as SiO and the melt back of a fraction of the seed. When the composition is "off eutectic" on the Si-rich side, the first portion of the boule to solidify directly under the seed usually shows extensive banding of the Si and eutectic regions. An example of banding in a section of the Si-TaSi$_2$ eutectic composite is shown in Figure 3. This initial section constitutes about 3% by volume of the boule. Banding in this section occurs primarily at the boule perimeter, denoted A in Figure 3(a). The higher magnification micrograph of region A in Figure 3(b) shows that the eutectic is very finely spaced, indicating that temperature fluctuations associated with the Czochralski growth have caused significant oscillations in the growth rate. The eutectic solidifies at the peak in the growth rate. A high magnification of the eutectic region, denoted B, is shown in Figure 3(c). This eutectic region shows a rod spacing that is much larger than occurs at the periphery of this boule section and is typical of the rod structure throughout the bulk of the boule. After about 5% by volume of the boule is solidified, the melt composition reaches the eutectic, banding becomes inconsequential, and the structure is more uniformly eutectic.

Banding seems to increase the frequency of nucleation of other Si grains or twins but does not preclude single-crystal growth. The boule in Figure 3 grew with a single-crystal

125

matrix from seed to tail. A photograph of a typical single-crystal matrix Si-TaSi$_2$ eutectic boule is shown in Figure 4. Single-crystal growth is possible despite a Si-rich composition, temperature fluctuations, and large variations in comparison. Single-crystal matrix eutectic growth is much more difficult when the melt is "off eutectic" on the metal-rich side. In this case, a mushy, two-phase zone is formed directly beneath the seed composed of melt and solid. Composite growth from a mushy zone invariably results in a polycrystalline Si matrix.

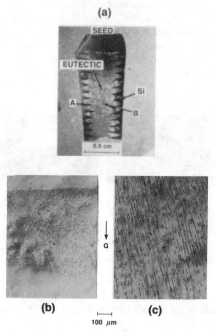

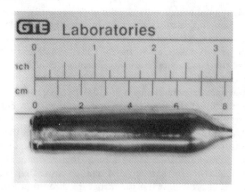

Figure 3.  (a) Banding of the Si-TaSi$_2$ eutectic just below the Si seed in a single-crystal Si matrix composite. V = 20 cm/h; (b) and (c) show an enlargement of the region denoted A and B respectively.

Figure 4.  A photograph of a single-crystal matrix Si-TaSi$_2$ eutectic composite boule, V = 20 cm/h.

Single-crystal matrix Si-silicide eutectics have only been grown in the $TaSi_2$- and $WSi_2$-containing systems. There is no a priori reason why with sufficient effort all the refractory metal disilicide eutectic systems having a small volume fraction of the second phase could not be grown with a single-phase semiconductor matrix. The higher volume fraction Si-silicide eutectics may be more difficult to achieve in a quasi-single-crystalline state.

### GaAs-Based Eutectics

Both GaAs-MoAs and GaAs-CrAs eutectics were grown by the LEC technique. Following growth the microstructures were examined in a similar way to the Si-based eutectics and compared to the Bridgman-grown structures obtained by Reiss and Renner (3). LEC growth of the GaAs-MoAs eutectic at a rate of 6 cm/h yielded an approximate interrod spacing of 5 $\mu$m. MoAs constituted about 6% by volume of the boule. It was difficult to obtain an accurate measure of the interrod spacing in this composite because the shape and distribution of the MoAs was quite nonuniform. Certain areas were dominated by a rod-like structure, while other areas contained mostly oriented plates. Still other regions of the boule showed irregular structures, with the MoAs forming tube-like particles. A micrograph of an area showing the three common shapes is shown in Figure 5(a). Reiss and Renner (3) did not report a similar nonuniformity in the Bridgman-grown eutectic.

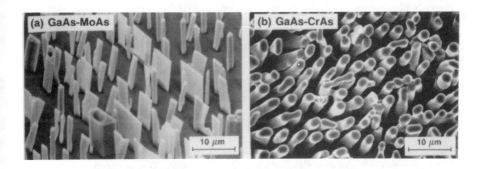

Figure 5. Scanning electron micrographs of two GaAs-based eutectics. The microstructure of the GaAs-MoAs eutectic is shown in (a) and the microstructure of the GaAs-CrAs eutectic is shown in (b). Both were pulled at 6 cm/h.

The LEC-grown GaAs-CrAs eutectic exhibited a more uniform microstructure throughout the boule. A typical region is shown in Figure 5(b). In this case, the 6 cm/h growth rate used yielded an average interrod spacing of 4.6 $\mu$m. The eutectic composition corresponds to 30 volume % CrAs. Relative to the Si-$CrSi_2$ eutectic, which similarly contains about 30 volume % metal phase, the GaAs-based system displays a more regular structure. Nevertheless, misalignment of the CrAs phase is evident in Figure 5(b), and rods are observed to intersect and form extended structures. The well-aligned rod structures apparent in a longitudinal section of the GaAs-CrAs-grown eutectic by Reis and Renner suggests that, in this case, the Bridgman-growth technique may yield a better rod alignment than the Czochralski technique.

## Discussion

Several different Si-metal silicide and GaAs-metal arsenide eutectics have been directionally solidified using the Czochralski crystal growth technique. The microstructure of the composites has been examined for suitability in a variety of electronic devices. For such applications, the microstructural features desired are a (1) rod-like structure, (2) small volume fraction of the metal phase (less than 10%), (3) regular distribution of rods, (4) interrod spacing less than about 10 $\mu$m, and (5) good alignment of the rods. Of all the Si-based systems, none appear to offer any microstructural advantage over the Si-TaSi$_2$ system. The Si-NbSi$_2$ and the Si-WSi$_2$ systems are structurally similar to the Si-TaSi$_2$ system and can probably be developed for similar applications. However, since the Schottky barriers of these *in-situ* metal/semiconductor junctions are expected to be comparable to that for the Si-TaSi$_2$ system (12), there is no advantage to be gained.

Unlike the Si systems, development of a GaAs-based system with a microstructure similar to the Si-TaSi$_2$ microstructure would be very desirable. Not only would the Schottky barrier be expected to be higher, but the GaAs matrix would also allow devices to operate faster and at higher voltages. Of the two systems examined, GaAs-MoAs is more similar to the Si-TaSi$_2$ system and meets all the microstructural requirements except for the uniformity of the rod distribution in the boule. Optimization of the growth technique to improve the boule uniformity is required. The GaAs-CrAs system has too high a volume fraction of the metallic phase but may be useful in certain applications if the alignment of the rods can be improved.

The only other documented SME GaAs-based composite is the GaAs-VAs system (3) Nevertheless, undiscovered SME systems containing metallic arsenides, gallium-based intermetallics, or even metal borides may exist. Evaluation of the phase diagrams of these systems may yield useful GaAs rod-like eutectic microstructures suitable for the development of eutectic devices.

## Acknowledgments

Discussions of eutectic growth with J. Gustafson and B. Yacobi are gratefully acknowledged. This work was sponsored in part by the SDIO/IST and managed by the Office of Naval Research and Contract No. N00014-86-C-0595. Additional support by the Air Force Office of Scientific Research under Contract No. F49620-86-C-0034 is noted.

## References

1. L.M. Levinson, "Highly anisotropic columnar structures in Si," *Appl. Phys. Lett. 21*, pp. 289–291 (1972).

2. N.J. Helbren and S.E.R. Hiscocks, "Silicon- and germanium-based eutectics," *J. Mater. Sci. 8*, pp. 1744–1750 (1973).

3. B. Reiss and T. Renner, "der gerichtete einbau von schwermetallphasen in GaAs," *Zeitschrift für Naturforshung 21*, pp. 546–548 (1966).

4. B.M. Ditchek and M. Levinson "Si-TaSi$_2$ in-situ junction eutectic composite diodes," *Appl. Phys. Lett. 49*, pp. 1656–1658 (1986).

5. B.G. Yacobi and B.M. Ditchek, "Characterization of multiple in-situ junctions in Si-TaSi$_2$ composites by charge collection electron microscopy," *Appl. Phys. Lett. 50*, pp. 1083–1085 (1987).

6. B.M. Ditchek, B.G. Yacobi, and M. Levinson, "Novel high quantum efficiency $Si$-$TaSi_2$ eutectic photodiodes," *Appl. Phys. Lett. 51*, pp. 267–269 (1987).

7. B.M. Ditchek, T.R. Middleton, P.G. Rossoni, and B.G. Yacobi, "A novel high voltage transistor fabricated using the in-situ junctions in a $Si$-$TaSi_2$ eutectic composite," submitted to *Appl. Phys. Lett.*

8. B.M. Ditchek, "Bridgman and Czochralski growth of $Ge$-$TiGe_2$ eutectic composites," *J. Cryst. Growth 75*, pp. 264–268 (1986).

9. M.C. Flemings, *Solidification Processing*, New York, NY: McGraw-Hill Book Co., pp. 93–113 (1974).

10. B.M. Ditchek, B.G. Yacobi, and M. Levinson, "Depletion zone limited transport in $Si$-$TaSi_2$ eutectic composites," *J. Appl. Phys.* (March 1988).

11. B.M. Ditchek, "Periodic $Si$-$CoSi_2$ eutectic structures," *J. Appl. Phys. 61*, pp. 5419–5424 (1987).

12. M.A. Nicolet and S.S. Lau, in *VLSI Electronics*, edited by N. Einspruch and G. Larrabee, Acadamic, New York, pp. 330 (1983).

# MULTIDIRECTIONAL
# SOLIDIFICATION

# MODELING OF EQUIAXED PRIMARY AND EUTECTIC SOLIDIFICATION

M. Rappaz,* and D.M. Stefanescu**

*Laboratoire de Metallurgie Physique
Ecole Polytechnique Federale de Lausanne
34, ch. Bellerive
CH-1007 Lausanne, Switzerland
**Department of Metallurgical Engineering
The University of Alabama
P.O. Box G
Tuscaloosa, AL   35487

ABSTRACT

The fraction of solid produced during primary and eutectic equiaxed solidification has been calculated using a number of models involving nucleation and growth laws for dendrites and euetctics equiaxed grains.  These subroutines have been incorporated in existing FDM programs, i.e., macro modeling of heat transfer has been coupled with micro modeling of equiaxed nucleation and growth.  This allowed for calculation not only of cooling curves or solidification isotherms, but also of microstructural features such as:   number of grains, microstructure spacings, fraction of eutectic, at the level of the whole casting.  Specific examples are shown for Al-Si and Fe-C alloys.

Solidification Processing of Eutectic Alloys
D.M. Stefanescu, G.J. Abbaschian and R.J. Bayuzick
The Metallurgical Society, 1988

# 1. Introduction

The topic of modeling of solidification of castings has received increased attention as the computer revolution matured. The main application of this technique has traditionally been to calculate the path of the isotherms through shaped castings. In turn, this can be used to predict the locations of hot spots in castings and thus check on the computer a proposed gating and risering system, rather than the classical trial and error technique used in foundries.

After the pioneering paper of Henzel and Keverian [1] describing the application of the Transient Heat Transfer program (an FDM program developed in 1959 at General Electric Co.) to solving heat transfer equations for castings, a variety of publications such as [2,3,4,5] to cite only a few, have dealt with the use of different numerical techniques in the area of solidification of castings. Although the next logical step forward, which is to include nucleation and growth kinetics in simulations, in order to generate information about the microstructure of the alloys and to correctly treat the heat generation problem, was envisioned by Oldfield as early as 1966 [6], the progress in this direction was slow to come. A followup paper by Stefanescu and Trufinescu [7] in 1974 seems to be the only one until 1984, although nucleation and growth kinetics was used in modeling cooling curves to interpret inoculation in cast iron [8].

Starting with 1984, a renewed interest in modeling microstructural evolution as part of simulation of casting solidification is manifested in the literature. An analytical approach was applied by Fredriksson and Svensson to solidification of eutectic gray, ductile and white iron [9], and then extended to hypoeutectic irons and to the eutectoid transformation by Stefanescu and Kanetkar [10]. Then an inner nodal direct FDM scheme combined with a model assuming growth of graphite spheroids controlled by diffusion of carbon through the austenite shell was used by Su et. al. [11]. More recently, developments along the same line have been made for dendrite growth of equiaxed grains [12-14].

The purpose of this paper is to discuss the state of the art of those techniques used for simulation of solidification of casting which include modeling of microstructural evolution.

## 2. Macroscopic Modeling

Solidification of alloys is primarily controlled by heat diffusion and to some extent by convection within the liquid region. In most approaches to solidification modeling of complex shape castings, the continuity equation of motion is not solved explicitly and correction is taken into account by increasing the heat conductivity above the melting point or the liquidus temperature. Under this assumption, the basic continuity equation which governs solidification at the macroscopic scale is that of conservation of energy. One has:

$$\text{div} \left( k(T) \cdot \vec{\text{grad}} \, T \, (\vec{x},t) \right) + \dot{Q} = \rho C_p(T) \frac{\partial T(\vec{x},t)}{\partial t} \tag{1}$$

where $T(\vec{x},t)$ is the temperature field, $k(T)$ the thermal conductivity, $\rho C_p(T)$ the volumic specific heat and $\dot{Q}$ the source term which is associated with the phase change. In solidification modeling, $\dot{Q}$ can be written as:

$$\dot{Q} = L\frac{\partial f_s(\vec{x},t)}{\partial t} \tag{2}$$

where $f_s(\vec{x},t)$ is the solid fraction and L the volumic latent heat.

In order to solve Equation (1), a relationship between the two fields, $T(\vec{x},t)$ and $f_s(\vec{x},t)$, must be found. As it will be shown in the next section for the case of equiaxed solidification, this can be done by considering the basic mechanisms of microstructure formation, i.e. nucleation and growth kinetics. A simple and widely used approach is to assume that the fraction of solid, $f_s$, depends only upon the temperature, T, and not upon cooling rate or growth rate. For pure metals or eutectic alloys, one can assume that $f_s = 0$ above the melting point or the eutectic temperature, and that $f_s = 1$ below the equilibrium temperature. For dendritic alloys, various models of solute diffusion have been developed [15]. They all assume complete mixing of solute within the liquid, thus resulting in a unique $f_s(T)$ curve.

Using the assumption that $f_s$ depends only upon T, Equations (1) and (2) can be combined to give:

$$\text{div}\left(k(T) \cdot \text{grad } T\ (\vec{x},t)\right) = \left(\rho C_p(T) - L\frac{\partial f_s}{\partial T}\right)\frac{\partial T(\vec{x},t)}{\partial t} \tag{3}$$

Defining the enthalpy, H, as:

$$H(T) = \int_0^T \rho C_p(\theta) \cdot d\theta - L(f_s(T)-1) \tag{4}$$

Equation 3 can also be written as:

$$\text{div}\left(k(T) \cdot \text{grad } T\ (\vec{x},t)\right) = \frac{\partial H(\vec{x},t)}{\partial t} \tag{5}$$

An effective specific heat, $\rho C_p^*$, can be derived from Equation 4:

$$\rho C_p^*(T) = \frac{\partial H}{\partial T} = \rho C_p(T) - L\frac{\partial f_s}{\partial T} \tag{6}$$

which, when introduced in Equation 3, gives:

$$\text{div}\left(k(T) \cdot \text{grad } T\ (\vec{x},t)\right) = \rho C_p^*(T) \cdot \frac{\partial T(\vec{x},t)}{\partial t} \tag{7}$$

The curves H(T) and $\rho C_p^*(T)$ which are shown in Figure 1 correspond to an Al-Si7% alloy. They have been calculated using a Brody-Flemings model of solute diffusion [15].

The most common numerical techniques which are used to solve the heat-flow equation (Eqs. 5 or 7) are the Finite Difference (FDM) and the Finite Element Methods (FEM) [2,3,4,5]. After enmeshing the space and using a time-stepping procedure, both methods finally give a set of non-linear equations which have to be solved at each time step. If an enthalpy method is used (Eq. 5), one has:

$$[K] \cdot \{T\} + \{bc\} = [M]\frac{\Delta\{H\}}{\Delta t} \tag{8}$$

135

Where [K] is the conductivity matrix, [M] the mass matrix, {T} the vector of temperatures at each node, Δ{H} the variations of enthalpy at each node and {bc} the vector corresponding to the boundary conditions. In the case of the specific-heat formulation (Eq.7), one gets:

$$[K] \cdot \{T\} + \{bc\} = [\rho C_p^*] \cdot \frac{\Delta \{T\}}{\Delta t} \qquad (9)$$

As it can be seen, the mass matrix [M] which is only related to the geometry of the enmeshment is now replaced by an effective specific-heat matrix, $[\rho C_p^*]$, which is temperature dependent. It should be pointed out that [M] and $[\rho C_p^*]$ have always a diagonal form when using FDM, while with FEM they have only a diagonal form if a mass-lumping technique is adopted.

A detailed comparison of the enthalpy method (Eq. 8) and of the effective specific-heat method (Eq. 9) is out of the scope of the present paper. The specific-heat method is certainly the most widely used technique as it can be easily implemented into standard FEM/FDM programs for thermal problems. However, some care has to be taken close to the discontinuities of the $\rho C_p^*(T)$ curve (see Fig. 1) if one does not want to miss the latent heat release. In practical applications, this means that a eutectic reaction has to be spread over a solidification interval, and the time steps chosen so as to have enough points within the solidification range. The enthalpy method is more difficult to implement but it has several advantages as mentioned in Ref. [16]. In particular, as the enthalpy is taken as the variable, conservation of energy is ensured no matter what are the time steps and the solidification interval. Furthermore, the enthalpy change is independent upon the solidification path, which represents also an advantage when dealing with macro-microscopic modeling.

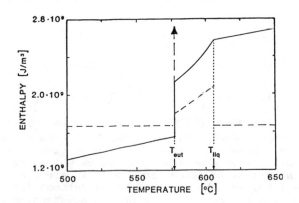

Figure 1
Enthalpy-temperature (full line) and effective specific heat-temperature (dotted line) relationships of an Al-7%Si alloy.

136

## 3. Microscopic Modeling

The macroscopic approach which has been briefly presented in the previous section can be reasonably applied to columnar solidification since the growth rate of the microstructure (eutectic front or dendrite tips) is more or less equal to the speed at which the corresponding isotherms move (eutectic or liquidus isolines). Therefore, microstructural parameters and undercooling can be directly calculated from the temperature field in this case.

When dealing with equiaxed microstructures, the growth speed of the grains is no longer related to the speed of the isotherms, but rather to the local undercooling. Furthermore, the solidification path is also dependent upon the number of grains which have been nucleated within the undercooled melt. In such a case, the approach which is undertaken has to relate the fraction which has solidified to the local undercooling. In a first approach, it is assumed that the bulk undercooling is very close to the interface (tip) undercooling.

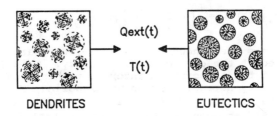

**DENDRITES**        **EUTECTICS**

Figure 2
Schematic view of equiaxed dendritic and
eutectic solidification.

### 3.1 Kinetics Approach

Let us consider a small volume element, $V$, of uniform temperature $T$, within which equiaxed solidification is proceeding (Fig. 2). At a given time, $t$, the fraction of solid, $f_s(t)$, is given by [17]:

$$f_s(t) = n(t) \cdot \tfrac{4}{3}\pi R^3(t) \cdot f_i(t) \qquad (10)$$

where $n(t)$ is the density of grains, $R(t)$ the average equiaxed grain radius characterizing the position of the dendrite tips or that of the eutectic front, and $f_i(t)$ the internal fraction of solid. For eutectics, the grains are fully solid and accordingly $f_i(t) = 1$ at any time. For dendritic alloys, $f_i(t)$ represents the fraction of the grains which is really solid.

In order to predict the evolution of the solid fraction, $f_s(t)$, one has to relate the 3 variables, $n(t)$, $R(t)$ and $f_i(t)$, to the undercooling, $\Delta T$. This can be done by considering nucleation, growth kinetics, and, for dendrites, solute diffusion.

## 3.2 Nucleation

Although the mechanisms of heterogeneous nucleation are not yet clearly known for most alloys, it is widely accepted that new solid grains form over a foreign substrate such as the surface of the mold or impurity particles. Based upon this mechanism, Turnbull [18] has shown that the rate, $\dot{n}(t)$, at which new grains are heterogeneously nucleated within the liquid can be given, at small undercooling, by:

$$\dot{n}(t) = K_1 \ (n_0 - n(t)) \ \exp \left\{ \frac{-K_2}{\Delta T(t)^2} \right\} \tag{11}$$

where $K_1$ is proportional to a collision frequency with nucleation sites, $n_0$ the total number of sites present in the melt before solidification and $K_2$ a constant related to the interfacial energy between substrate and nucleated grain. Since the constants, $K_1$, $n_0$ and $K_2$ are almost impossible to predict, they have to be deduced from experiment. Once they are known, the grain density, $n(t)$, can be predicted at each time by integrating Eq. 11 over time or temperature:

$$n(t) = \int_{t_0}^{t} \dot{n}(\tau) d\tau = \int_{0}^{\Delta T(t)} \dot{n}(T) \ . \ \frac{dT}{\left(\frac{dT}{dt}\right)} \tag{12}$$

However, it has been shown recently [19,20] that this approach fails to predict the correct grain density. This is related to the fact that the temperature interval within which nucleation proceeds is very narrow. For an undercooling, $\Delta T$, smaller than a critical value, $\Delta T_N = \sqrt{K_2}$, there is no significant nucleation, and as soon as $\Delta T_N$ is reached, $n(t)$ increases very rapidly to its saturation limit, $n_0$ (see Fig. 3 and 5). For that reason, the complex nucleation law of Eq. 11 can be almost replaced in solidification modeling by a Dirac function

$$\frac{dn}{dT} = n_0 \ . \ \delta(T-T_N) = n_0 \ . \ \delta\left(\Delta T - \sqrt{K_2}\right) \tag{13}$$

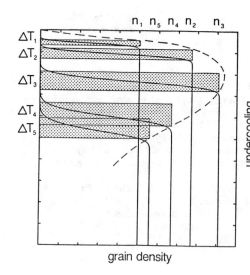

$n_1 \ \ n_5 \ \ n_4 \ \ n_2 \ \ \ \ \ n_3$

$\Delta T_1$
$\Delta T_2$
$\Delta T_3$
$\Delta T_4$
$\Delta T_5$

undercooling

grain density

Figure 3
Schematic illustration of heterogeneous nucleation occurring on a family of inoculant sites, characterized by a density of sites, $n_{0i}$, and by a critical temperature, $T_{Ni}$, at which nucleation occurs [20].

If more than one type of nucleation sites are present, one can introduce a set of Dirac functions (see Fig. 3):

$$\frac{dn}{dT} = \sum_i n_{o,i} \, \delta(T-T_{N,i}) \tag{14}$$

This discrete distribution of nucleation site types can be replaced as well by a continuous distribution (figure 4). Although this last approach may not reflect the complex phenomena of heterogeneous nucleation, it has several advantages in microscopic modeling of solidification [20]: i) the number of adjustable parameters is limited to 2 or 3; ii) the grain density, n, is now only function of the undercooling, i.e., the time-dependence has been removed; iii) nucleation stops as soon as recalescence occurs; iv) the grain density increases with the cooling rate, $|\dot{T}|$, as the minimum of the cooling curve is lowered with increasing values of $|\dot{T}|$; and v) the parameters of the nucleation distribution can be easily derived from experiment.

In fact, a continuous distribution of nucleation site types can be replaced by a very narrow distribution if one only wants to simulate heterogeneous nucleation occurring at a given undercooling, $\Delta T_N$, with a given density of sites, $n_o$ (Eq. 13). This last approach can be used for eutectic solidification based on the fact that, as previously discussed, the nucleation interval is very narrow. E.g., for cast iron it was calculated to be of about $0.1^{\circ}C$ (see Fig. 5) [19]. Thus a nucleation temperature, $\Delta T_N$, at which all eutectic grains nucleate at the same time is chosen. For the case of irons with non-uniform grains, it must be assumed that different types of substrates become active at different nucleation temperatures. Accordingly, several nucleation temperatures need to be selected, at which fractions of the final number of nuclei are generated.

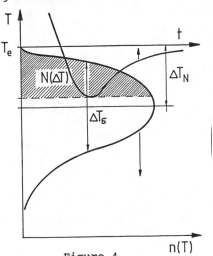

Figure 4
Continuous distribution of nucleation site types. [17,20]

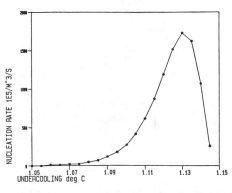

Figure 5
Calculated relationship between nucleation rate and undercooling in cast iron [19].

## 3.3 Kinetics and Solute Diffusion

Evolution of the grain radius, $R(t)$, can also be related to the undercooling, $\Delta T$, of the volume element (it should be noted that only constitutional undercooling is being considered here, as kinetics is primarily controlled by solute diffusion in metallic alloys). Jackson and Hunt [21] have shown that the speed of a eutectic front, $v$, is related to the undercooling through the relationship:

$$v = \frac{dR}{dt} = \mu \cdot (\Delta T)^2 \qquad (15)$$

where $\mu$ is a constant depending upon the characteristics of the alloy.

For dendritic alloys, Esaka and Kurz have deduced a similar relationship in the approximation of low Péclet number [22]: it also relates the square of the undercooling, $\Delta T$, to the velocity $v$, of the dendrite tips, Therefore, equation 15, with a different $\mu$ value, can be used to predict the evolution of the grain size. However, in this case, one has still to calculate the evolution of the internal volume fraction of solid, $f_i(t)$ (Eq. 10). For that purpose, Rappaz and Thévoz [23] have recently developed a solute diffusion model which is summarized in Figure 6. Assuming that there is complete mixing of solute within the interdendritic liquid of the spherical

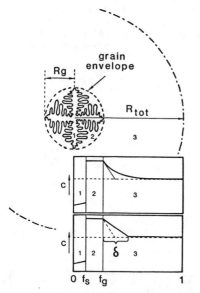

grain
envelope

$R_g$

$R_{tot}$

a

b

c

$0\ f_s\ f_g$   1

Figure 6

A schematic view of the solute diffusion model developed for equiaxed dendritic growth.
a) The 3 regions which can be distinguished are: 1) solid dendrite; 2) interdentritic liquid where complete mixing of solute is assumed; 3) liquid outside the grain envelope where diffusion occurs; b,c) concentration profiles corresponding to (a), according to Refs [14] and [23], respectively.

grain envelope outlined by the dendrite tip position, they considered the solute balance at the scale of the equiaxed grain and the solute flow leaving out the grain envelope. They found that:

$$f_i(t) = \Omega(t) \cdot g\ (\delta, R) \qquad (16)$$

where $\Omega = \dfrac{C^* - C_0}{C^*\ (1-k)}$ is the supersaturation, and $g(\delta, R)$ a correction function which takes into account the solute layer, δ, around the grain envelope, $C^*$ is the concentration within the interdendritic liquid (see Fig. 6), $C_0$ the initial concentration and k the partition coefficient. Since the undercooling $\Delta T$ is equal to $m(C^*-C_0)$, where m is the slope of the liquidus, $f_i(t)$ is again directly related to $\Delta T$ through Eq. 16.

From the solute flux balance, it has then been shown that the solute layer, δ, is simply given by the ratio $2D/v$, where D is the diffusion coefficient. The effect of δ on solidification is most noticed when the solute layers of neighboring dendritic grains overlap, thus changing the concentration $C_0$ which appears in the supersaturation.

## 3.4 Grain Impingement

Equation 10 assumes that the grains are spherical during the whole solidification process, and therefore that they do not impinge on each other. For dendritic alloys, the diffusion layer, δ, outside of the grain envelope, R, somehow takes already into account grain impingement. For eutectic grains, grain impingement can be introduced as follows: if the effective solid-liquid interface is labeled, S, then the increment of solid fraction, $df_s$, is given by:

$$df_s = S \cdot dR/V \tag{17}$$

where dR is the increment of solid normal to the interface, and V the total volume. Considering still spherical grains, one can write that:

$$S = nV4\pi R^2 \cdot F \tag{18}$$

The factor F whose value ranges between 0 and 1 tells which portion of the surface of the spherical grains is still in contact with the liquid.

Avrami [24] and Johnson-Mehl [25] correction for grain impingement predicts that:

$$F = 1-f_s \tag{19}$$

from which one deduces, by integration of Eq. 17:

$$f_s = 1 - \exp \left( -n \cdot \frac{4}{3}\pi R^3 \right) \tag{20}$$

Equation 20, which was originally developed for solid state transformation, does not take into account the fact that equiaxed grains can move within the liquid. Furthermore, the radius R must go to infinity in order to reach $f_s = 1$ (see Fig. 7). For those reasons, geometrical modeling of grain impingement has been considered recently to calculate the factor F appearing in Eq. 18 [20]. Based upon a close-pack model of grains of equal radius R, the relationship $f_s(R)$ plotted in Figure 7 can be derived. As it can be seen, it deviates substantially from the standard Johnson-Mehl or Avrami correction.

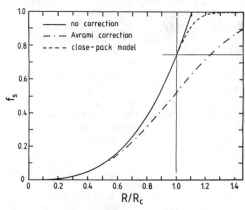

Figure 7
Grain impingement models relating the solid fraction, $f_s$, to the average grain radius, R. (The grain radius R has been normalized by $R_c$, a critical radius which corresponds to $f_s$ = 0.74, i.e. to the solid fraction of densely packed spheres).

## 3.5 Time-stepping Procedure

Assuming that the heat-flow leaving out the specimen is known at any time (see Fig. 2), solidification of equiaxed microstructures can be predicted according to the following time-stepping calculation:

i) knowing the temperature at time T, and thus the undercooling $\Delta T$, the grain density is updated from the previous time-step according to the nucleation distribution (see Fig. 4):

$$n(t) = n(t-\Delta t) + \Delta n \qquad (21)$$

ii) the new grain radius, $R_{new}$, is calculated so that the fraction of solid remains unchanged while step (i) is being done:

$$n(t) \cdot R^3_{new} = n(t-\Delta t) R^3_{old} + \Delta n R^3_N \qquad (22)$$

($R_N$ is the radius of the new nuclei. It can be set to 0).

iii) from the undercooling, $\Delta T(t)$, the growth speed, v, of the grain is deduced (Eq. 15). For eutectics, the variation of $\Delta f_s$ between t and t + $\Delta t$ can be readily calculated using Eqs. 17 and 18:

$$\Delta f_s = f_s(t+\Delta t) - f_s(t) = n(t) \cdot 4\pi R^2(t) \cdot F(f_s(t)) \cdot v \cdot \Delta t \qquad (23)$$

For dendritic alloys, the internal volume fraction of solid, $f_i(t)$, is known and one has:

$$\Delta f_s = n(t) \cdot 4\pi R^2(t) \cdot f_i(t) \cdot v\Delta t + n(t) \cdot \frac{4}{3}\pi R^3(t) \cdot \Delta f_i(t) \qquad (24)$$

The first term which appears at the right of Eq. 24 is known, while the second one is directly related to the change of temperature between time t and t+$\Delta t$ (see Eq. 16). It can be grouped with the specific heat term in Eq. 25 (see Ref. [27]).

iv) The variation of temperature between t and (t+Δt) is computed according to the heat balance:

$$\frac{Q_{ext}}{V} \cdot \Delta t = \rho C_p \left[ T(t+\Delta t) - T(t) \right] - L \, \Delta f_s \qquad (25)$$

where $Q_{ext}$ is the heat leaving the specimen per unit time. $\Delta f_s$ is taken from Eq. 23 or from Eq. 24.

v) knowing the new temperature, $T(t+\Delta t)$, the calculation can restart at point (i) until solidification is finished.


4.  Macro-microscopic Modeling

When dealing with modeling of equiaxed solidification in the previous section, it was assumed that the heat, $Q_{ext}$, coming out of the volume V (see Fig. 2, Eq. 25) was known. If we now consider that this volume element is part of the enmeshment of a whole casting, $Q_{ext}$ has to be calculated for each mesh from the continuity of energy (section 2). The coupling between the macroscopic heat flow equation and the microscopic models of solidification can be achieved according to various schemes. Two basic coupling schemes are shown in Fig. 8.

The "latent-heat method" [19] shown in Fig. 8a is the most straightforward one. Formulating Eq. 3 with FDM or FEM, the variations $\Delta\{f_s\}$ at all nodes are calculated according to the microscopic model of solidification (Eqs. 23 or 24). In the case of eutectics, $\Delta\{f_s\}$ is calculated from the undercooling at each node at time t (Eq. 23), and thus this is a known vector which can be readily integrated into the heat-flow equation in order to deduce the variations of temperature, $\Delta\{T\}$, at all nodes. For dendritic alloys, $\Delta\{f_s\}$ is the sum of a known contribution related to the velocity of the dendrite tips and a contribution of the variation of the internal volume fraction of solid, $\Delta\{f_i\}$, which can be grouped with the specific heat contribution [23] (see Eq. 24). In both dendritic or eutectic alloys, the variation $\Delta\{T\}$ can be derived explicitly or implicitly while the variation $\Delta\{f_s\}$ is given in an explicit fashion by the undercoolings at each node at time t. For example, in the EUCAST program (an FDM program developed at The University of Alabama) the latent heat term was calculated at time $t_i$ using values of $\{T\}$ at time $t_i$ [19,28]. This resulted in time step dependency of $\{\Delta T\}$. To avoid this problem, in the BAMACAST program (a volume control method program developed at The University of Alabama) a predictive-corrective method was used. The latent heat term was calculated implicitly by linearizing the $\{\Delta H\}$ term (known $\Delta H$ for $t_i$ was extrapolated to unknown $\Delta H$ for time $t_{i-1}$) [23].

The second scheme presented on Fig. 8b has been implemented into the 3-MOS program (a FEM code developed at EPFL from the library Modulef [16]). It is based upon an enthalpy method [27]. Since the variation of enthalpy is independent upon the solidification path once the heat-flow is known, the macro- and the microscopic calculations can be somehow decoupled. At the macro level, one can still solve the heat-flow equation as mentioned in section 2. Once the variations of enthalpy, $\Delta\{H\}$,

at all nodes are known, the solidification path can be computed
taking into account that the variation $\Delta H_j$ at a given node,
j, is nothing but the left-hand term of Eq. 25. As can be seen
in Fig. 8b, the macroscopic time-step, $\Delta t$, can be subdivided
into many smaller time-steps, $\delta t$, to perform the microscopic
calculations, assuming that heat removal is made at constant
rate during $\Delta t$. The only constraint set on $\Delta t$ is coming now
from the heat-flow and not from energy balance: if $\Delta t$ is too
large, the heat-flow, which is computed at the macroscopic
scale, may not be correct as a result of recalescence of some
nodes, for example. This may result in cooling curves of

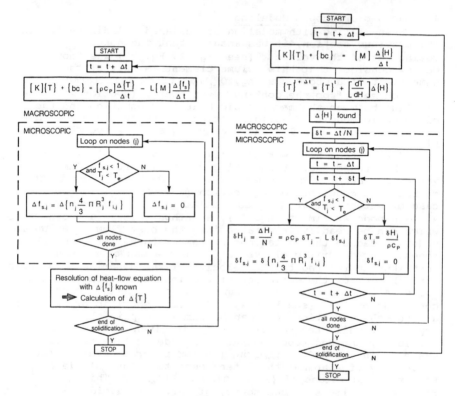

Figure 8
Flow charts of macroscopic-microscopic modeling of
solidification based upon two different schemes:
a) latent heat method; b) enthalpy method.

neighboring nodes which intersect during recalescence. Although
this result is contrary to physics, the macro-microscopic
coupling based upon the enthalpy scheme seems to give better
convergence of the calculated values.

5.  Some Results
    The results which are shown hereafter illustrate the
possibilities of integrating microscopic modeling of
solidification into macroscopic heat-flow calculations.

Figure 9 shows the recalescences of two Al-7%Si specimen
[23]. The dotted curves have been measured at the center of two
small volumes containing the alloy. Curve B corresponds to the
uninoculated melt and the final grain size was about 2mm in
radius. When adding 50 ppm of Ti prior to solidification
(curve A), the final grain radius was reduced to 0.5mm. The
continuous curves shown in Fig. 9 have been computed with the
analytical model of solute diffusion and based upon the measured
grain sizes. The heat leaving out of the specimen (see Eq. 25)
was assumed to be constant and was deduced from the cooling rate
above the liquidus. Although the agreement between experiment
and modeling is excellent as far as depth and width of the
recalescence are concerned, it should be mentioned that the
kinetics law of the free dendrite tip had to be modified for
both specimens by a constant factor [23].

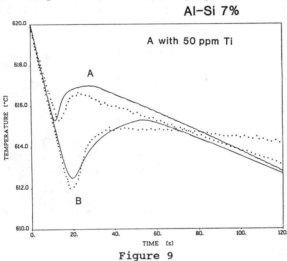

Figure 9
Measured (dotted lines) and calculated (solid lines)
recalescences for two Al-7%Si alloys. a) with 50 ppm
of Ti (final grain radius 0.5 mm); b) without inoculant
(final grain radius 2mm) [23].

The 6 cooling curves which are shown in Fig. 10 have been
measured for a one-dimensional gray cast iron casting [27]. A
3%C, 2.5%Si iron melt was poured in a ceramic mold over a copper
chill plate. 6 thermocouples were placed at various distances
from the chill plate (10/20/40/60/80/100 mm). Thermocouples
soldered within the copper chill permitted to deduce the
heat-flow entering the plate or leaving out the casting. This
flow was then used as the boundary condition to calculate the
corresponding cooling curves with the macro-microscopic model of
solidification. The effect of Si upon the mechanism of eutectic
growth was taken into account by modifying the equilibrium
eutectic temperature according to a Scheil model of Si
segregation. The initial temperature used in the calculation
($1350^0$C) corresponds to the value measured within the melt
just before pouring. Due to the time response of the
thermocouples and to the filling time, the measured curves start

145

at about 1320°C. Although the agreement between modeling and
experiment is poor in the liquid region (above 1160°C),
solidification is very well predicted with the macro-micro
model. In particular, calculated recalescence undercooling and
end of solidification are in good agreement with the
experimental curves. One can notice, however, that
solidification of primary phase close to 1190°C was not
included in the modeling.

Using the BAMACAST program for prediction of cooling curves
for gray iron [26] resulted in a close match with the
experimental results for cooling rates and undercooling, as
shown in Fig. 11. Details on nucleation and growth constants
used in the model are given in reference [28]. The only
remarkable discrepancy is at the end of solidification, probably
due to silicon segregation, which was not taken into account by
the model. A typical computer output, showing both the progress

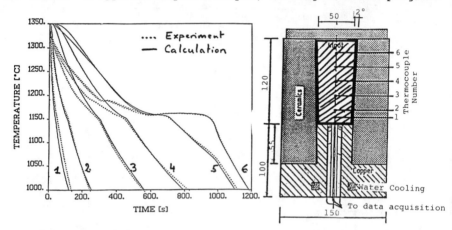

Figure 10
Measured (dotted lines) and calculated (solid lines)
cooling curves for cast iron poured over a copper chill
plate [27]. Height of the castings: 120mm; number of
meshes: 120. The parameters of nucleation deduced from
separate micro-castings experiments are the following:
gaussian distribution, center at 20K undercooling,
standard deviation: 4.75K, and total density of sites:
$1.2 \cdot 10^{11} m^{-3}$.

of solidification and the width of the mushy zone is given in
Fig. 12. Note the non-linearity of the fraction of solid with
respect to time and distance.

Theoretical predictions of the width of the mushy zone are
also in good agreement with experimental measurements. This is
shown in Fig. 13 where the open dots or squares are experimental
values for a cast iron sample. The beginning and end of
solidification for the two experimental points have been
determined from the cooling curve, by using computer aided
cooling curve analysis.

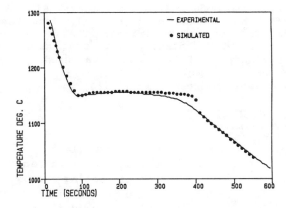

Figure 11
Calculated and experimental cooling curves for eutectic gray iron poured in a 5cm diameter bar molded in pep set sand. Thermocouples were inserted in the middle of the casting [26].

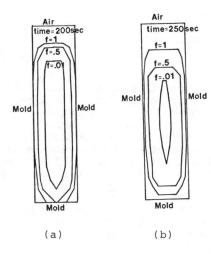

(a)　　　　　(b)

Figure 12
Calculated progression of the solidification front for a 5cm diameter cast iron bar poured in pep set sand mold [26]: a) after 200 seconds; b) after 250 seconds.

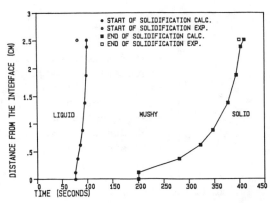

Figure 13
Calculated start and end of solidification wave fronts for a 5cm diameter bar and experimental points for a thermocouple placed at the center of the bar [26].

Good predictions were also obtained for eutectic Al-Si alloys poured in two different diameter bars (Fig. 14). Because the data base for Al-Si alloys is rather limited, it was necessary to calculate the growth constant from theoretical principles. A growth constant, $\mu$, of $5 \times 10^{-7} m/s.C^2$ was used, together with number of nuclei of $3 \times 10^9 /m^3$ and $1.3 \times 10^9 /m^3$ for bars of 3cm and 5cm diameters, respectively [26].

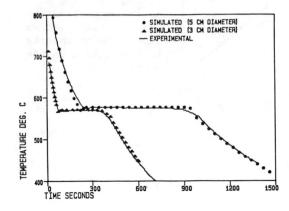

Figure 14
Calculated and experimental cooling curves for eutectic Al-Si alloys poured in pep set sand molds [26].

One of the main interest of macro-microscopic modeling of solidification is to predict microstructural features and not only cooling curves. Figure 15 compares the grain radius measured and calculated at the 6 locations of the thermocouples where the cooling curves shown in Figure 10 are recorded. (These radii are plotted as a function of the distance from the copper chill plate). The distribution of nucleation sites (see Fig. 4) was a gaussian line shape whose parameters were deduced from micro-castings of the same alloy. Although the discrepancy between experiment and modeling may be substantially large (especially for thermocouple #5), the trend of increasing the grain size with increasing distances from the chill (or decreasing cooling rates) is correctly predicted.

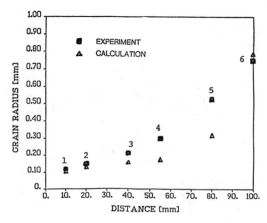

Figure 15
Measured (squares) and calculated (triangles) grain radii at the location of the thermocouples which recorded the cooling curves shown in Fig. 10 [27].

148

Another application of macro-micro modeling in the field of microstructure evolution is to predict the gray/white transition in cast iron [29]. For the general case of cast iron, two different eutectics must be considered: the stable one, $T_{st}$, and the metastable one, $T_{met}$. If then solidification occurs above $T_{met}$, the iron is gray, while if it occurs under $T_{met}$ the iron is white. The EUCAST program has been used to predict the gray/white transition by incorporating a white subroutine which calculates the solidification of cast iron using different number of nuclei and growth constant, as soon as the running temperature T drops below $T_{met}$, when solidification is not yet completed (f<1). Thus, the program can be used in two different ways: i) to calculate the amount of gray and white structure in a bar of a given diameter and ii) to calculate the diameter for which the structure becomes completely white for a given mold material (cooling rate), i.e. the gray/white transition.

Theoretical curves which can be used to calculate the amount of white and gray structure within a solidifying bar are given in Fig. 16. A total of 10 nodes were used (node 1 in the center, node 10 at the edge of the casting). It can be seen that starting with node 8, solidification occurs under $T_{met}$. This means that the last three nodes (8, 9 and 10) which is 30% of the sample starting from the outside, solidified white,while the 70% center part solidfied gray [28].

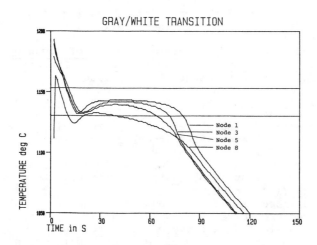

Figure 16
Theoretical cooling curves for different nodes
(shells) of a 2.2cm diameter bar [29].

149

# 5. Conclusions

It is quite obvious that macro-micro modeling of equiaxed primary and eutectic solidification can be performed satisfactorily at the present stage. Two particular areas need special attention in order to allow for the use of existing programs in a predictive mode: new heterogeneous nucleation models and data base for various alloys for both nucleation and growth of grains.

Work is already underway in extending this approach to solid state transformations, after which it will be just one additional step to predict mechanical properties in shaped castings.

## Acknowledgements

The Swiss project is supported by the Office Federal de l'Education et de la Science, Switzerland, and the American project is supported by NSF - EPSCoR in Alabama.

## REFERENCES

1. J.G. Henzel, Jr. and J. Keverian, "Comparison of Calculated and Measured Solidification Patterns for a Variety of Steel Castings", AFS Trans. (1965) 661-672.
2. R.D. Pehlke, R.E. Marrone and J.O. Wilkes, Computer Simulation of Solidification (American Foundrymen's Society, Des Plaines, Illinois, 1976).
3. H.D. Brody and D. Apelian, ed., Modeling of Castings and Welding Processes (The Metallurgical Society of AIME, 1981).
4. J.A. Dantzig and J.T. Berry, ed., Modeling of Casting and Welding Processes II (The Metallurgical Society of AIME, 1984).
5. H. Fredricksson, ed., State of the Art of Computer Simulation of Casting and Solidification Processes (Les Editions de Physique, France, 1986).
6. W. Oldfield, "A Quantitative Approach to Casting Solidification: Freezing of Cast Iron", ASM Trans, 59 (1966) 945-959.
7. D.M. Stefanescu and S. Trufinescu, "Zur Kristallisation-skinetik von Grauguss", Zeitschrift Metallkde, 65 (9) (1974) 610-66.
8. O. Yanagisawa and M. Maruyama, "Silicon Inoculation Mechanism in Cast Iron", 46th International Foundry Congress (1979) paper 21.
9. H. Fredriksson and I.L.. Svensson, "Computer Simulation of the Structure Formed During Solidification of Cast Iron", in The Physical Metallurgy of Cast Iron, H. Fredriksson and M. Hillert, ed. (North Holland, New York, 1984) 273-284.
10. D. M. Stefanescu and C. Kanetkar, "Computer Modeling of the Solidification of Eutectic Alloys: The Case of Cast Iron", in Computer Simulation of Microstructural Evolution, D.J. Srolovitz, ed. (The Metallurgical Society, Warrendale, Pa., 1985), 171-188.
11. K.C. Su, I. Ohnaka, I. Yaunauchi and T. Fukusako, "Computer Simulation of Solidification of Nodular Cast Iron", in

The Physical Metallurgy of Cast Iron, H. Fredriksson
and M. Hillert, ed., (North Holland, New York, 1984)
181-189.
12. I. Dustin and W. Kurz, "Modeling of Cooling Curves and
Microstructures During Equiaxed Dendritic Solidification",
Zeitschrift für Metallkunde 77 (1986) 265.
13. S.C. Flood and J.D. Hunt, "Columnar and Equiaxed Growth I
and II", Journal of Crystal Gorwth 82 (1987) 543/552.
14. M. Rappaz and P. Thévoz, "Solute Duffusion Model for
Equiaxed Dendritic Growth", Acta. Met. 353 (1987)
1487.
15. W. Kurz and D.J. Fisher, Fundamentals of
Solidification, (Trans. Tech. Publ., Aedermanssdorf,
Switzerland, 1986).
16. J.L. Desbiolles, M. Rappaz, J.J. Droux and J. Rappaz,
"Simulation of Solidification of Alloys Using the FEM Code
Modulef", in Ref. 5, 49-55.
17. M. Rappaz, Ph. Thévoz, Zou Jie, J.P. Gabathuler and H.
Lindscheid, "Micro-macroscopic Modeling of Equiaxed
Solidification", in Ref. 5, 277-284.
18. D. Turnbull, "Kinetics of Heterogeneous Nucleation",
J. Chem. Phys. 18 (1950) 198.
19. D.M. Stefanescu and C. Kanetkar, "Computer Modeling of the
Solidification of Eutectic Alloys: Comparison of Various
Models for Eutectic Growth of Cast Iron", in Ref. 5,
255-266.
20. Zou Jie, P. Thévoz and M. Rappaz, private communication,
1987.
21. K.A. Jackson and J.D. Hunt, "Lamellar and Rod Eutectic
Growth", Trans. Met Soc. of Am. Inst. Min. Engr., 236
(1966) 1129-1142.
22. H. Esaka and W. Kurz, "Columnar Dendrite Growth:  A
Comparison of Theory", J. Cryst. Growth, 69 (1984)
362.
23. M. Rappaz and Ph. Thévoz, "Analytical Model of Equiaxed
Dendritic Solidification", in Solidification Processing
1987 (Sheffield, UK), H. Jones, Ed.
24. M. Avrami, J. Chem. Phys. 8 (1940) 212.
25. W. A.Johnson, and  R.F. Mehl "Reaction Kinetics in
Processes of      Nucleation and Growth," (AIME Tech Pub.,
1089, 1939), 5.
26. C.S. Kanetkar, D.M. Stefanescu, N. El-Kaddah and I.G. Chen,
"Macro-Microscopic Simulation of Equiaxed Solidification of
Eutectic and Off-Eutectic Alloys", in Solidification
Processing 1987, (Sheffield, UK) H. Jones, Ed.
27. Ph. Thévoz, Zou Jie and M. Rappaz, "Modeling of Equiaxed
Dendritic and Eutectic Solidification in Castings", in
Solidification Processing, 1987, (Sheffield, UK), H.
Jones, ed.
28. D.M. Stefanescu and C.S. Kanetkar, "Modeling of
Microstructural Evolution of Cast Iron and Aluminum-Silicon
Alloys" in 54th International Foundry Congress
(New Delhi, India, 1987), U.S. Exchange Paper.
29. D.M. Stefanescu and C.S. Kanetkar, "Modeling of
Microstructural Evolution of Eutectic Cast Iron and of the
Gray/White Transition", Trans. Am. Found. Soc. 95
(1987) paper 68.

A COMPARISON BETWEEN THE GROWTH PROCESSES OF CAST IRON BY

THERMAL ANALYSIS

H. Fredriksson and I.L. Svensson

Department of Casting of Metals
The Royal Institute of Technology
S-100 44 Stockholm, Sweden

Abstract

Cast iron alloys can produce different solidification structures, containing flake like graphite, undercooled graphite, vermicular graphite, nodular graphite and cementite. The resulting structure is depending on the eutectic reaction. A theoretical analysis of the different types of eutectic reactions occuring during the solidification process is performed. It is found that the shape of the cooling curve varies with the type of structure formed due to differences in kinetic between the various eutectic reaction.

Solidification Processing of Eutectic Alloys
D.M. Stefanescu, G.J. Abbaschian and R.J. Bayuzick
The Metallurgical Society, 1988

153

## Introduction

During the solidification process of cast iron castings a number of substructures will be formed. The solidification starts with a primary precipitation of austenite in a dendritic pattern. During the continuing precipitation process eutectic structures will be formed. These structures can mainly be divided into five different types, namely: flake graphite, undercooled graphite, nodular graphite, vermicular graphite and white structure. The amount as well as the coarsness of each substructure is important for the properties. During the last years different types of computer models have been derived in order to predict the structure (1-3). In this paper we will extend those models and make calculations of the solidification process for all different structures formed. In the first part of this paper we will discuss the precipitation of austenite and there upon discuss the cooling curves formed at different types of eutectic reactions. The aim is to see in what range the cooling curves can be used to predict the type of structure formed. The calculations are performed on binary Fe-C alloys.

## Theoretical analysis

When the liquid comes into the mould the liquid is cooled by the mould. The heat transport for cooling the castings is normally very complex and is nowadays described by different types of computer models. In our case we will use the simplified method first derived by Chworinoff (4). The expression used is presented in annex 1.

The solidification starts with precipitation of austenite. To describe this precipitation we are going to use the model first derived by Fredriksson and Olsson (5). The necleation of austenite crystals starts randomly in the liquid and the crystals are moving around in the liquid, due to the convection. They collide with each other, and a network of dendrites are formed. Equations 2-7 in annex 2 describe the growth process of the crystals. The first term on the right hand side in equation (2) describes the heat released by the growing crystals and the second term describes the changing of the volume fraction due to temperature changes inside the crystals already formed. The growth rate of the crystals and the volume fraction of solid inside the crystals are described by equations (5) and (6). The concentration of carbon in the liquid as well as in the solid as function of temperature is described by the binary Fe-C phase diagram.

Figure 1 shows a calculation performed for an alloy with 3.8%C. The calculations are performed with the constants given in table 1. The figure shows that one plateau is apparent at the start of precipitation of austenite and another plateau occurs when the eutectic reaction starts. From the cooling curve the volume fraction of primary austenite can be evaluated as well as the dendritic coarsness. The eutectic reaction calculated in figure 1 will be discussed further on in this paper.

Table 1. Values of symbols

$Q$ = entalpy $\langle J \rangle$

$K_f$ = heat conductivity of mould, 0.7 $\langle J/s,M,K \rangle$

$C_p^f$ = heat capacity of mould, 700 $\langle J/kg,K \rangle$

$\rho_f$ = density of the mould 1500 $\langle kg/M^3 \rangle$

$\rho_m$ = density of the metal 7000 $\langle kg/m^3 \rangle$

$t$ = time $\langle s \rangle$

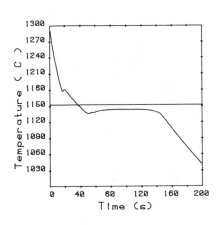

Figure 1.
Cooling curve with a primary precipitation of austenite followed by an eutectic reaction. The constants used are given in table 1 and annex 2. The cooling rate is that of a cylinder with a diameter of 20 mm.

$T_0$ = room temperature, 20 <C>

$T_c$ = casting temperature 1300 <C>

$T_i$ = solidification temperature, 1153 <C>

$A_i$ = surface area of casting $6.28 \cdot 10^{-2}$ <m²>

$V_0$ = volume of casting $3.14 \cdot 10^{-4}$ <m³>

$C_p^m$ = heat capacity of liquid, 700 <J/kg, K>

$\Delta H$ = heat of fusion, 220000 <J/kg>

$T_L^\gamma$ = equilibrium temperature liquid/austenite

$\sigma$ = surface tension (austenite/graphite) 0.2<J/m²> ($\gamma$/cem)0.8<J/m²>

$Ng$ = number of eutectic cells $4 \cdot 10^{10}$<N/m³>(otherwise stated in the figure)

The eutectic reaction in cast iron alloys is very complex and is influenced by a number of factors. It is well-known that we distinguish between flake graphite and undercooled graphite. It has several times been discussed that those two morphologies grow by different growth mechanism (1, 6). It has been proposed that undercooled graphite grows by a normal eutectic reaction and that flake graphite grows through a divorced eutectic reaction. However, it is difficut to theoretically describe the differences in reaction. Therefore we will use one growth law for the two morphologies. We will assume that graphite is formed by a normal eutectic reaction. Using the model for eutectic growth derived by Fredriksson (7) one gets the equations presented in appendix 3. In our treatment we will desregard the effect of the interface kinetic. By using the data given in table 1 and by combining the equations given in annex 1 and 3, cooling curves have been calculated. The result is shown in figure 2. The figure shows that one gets a curve with some recalecence followed by a solidification process with a fairly constant undercooling. From the curve two characteristic parameters can be evaluated, namely $\Delta T_{max}$ and $\Delta T_{min}$ defined in figure 2. Those two values are very much influenced by the number of growing units. A series of calculations were performed and those two parameters were evaluated as function of the number of growing units. The result is shown in figure 3.

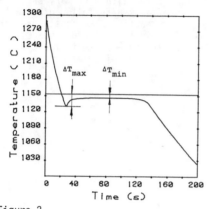

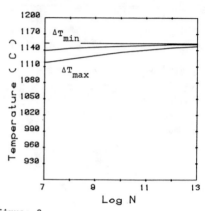

Figure 2.
Cooling curve for flake like graphite.

Figure 3.
Minimum and maximum undercooling as function of cell numbers for flake like graphite.

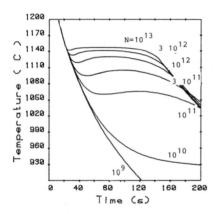

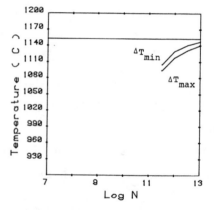

Figure 4.
Cooling curves for nodular cast iron calculated for different numbers of growing nodules.

Figure 5.
Minimum and maximum undercooling as function of cell numbers for nodular cast iron,

The growth process of nodular cast iron is very different from the growth process of flake graphite. The process is described in annex 4. By combining the equations given in annex 1 and 4 the cooling curves were calculated. The result is shown in figure 4. In the figure a series of cooling curves with different numbers of growing nodules is shown. Also in this case $\Delta T_{max}$ and $\Delta T_{min}$ was derived, the result is shown in figure 5. A comparison with figures 2 and 3 show that the undercooling is much more sensitive for the number of growing units for nodular cast iron than for flakelike cast iron. At very high numbers of growing units the curves look very similar. However this is with so many number of growing units that it is doubtful if it ever will exist.

The growth process of vermicular graphite is the most difficult one to describe of all the substructures formed. There are no reports describing the kinetics. In our case we will use the model described in annex 5. We have chosen a linear growth law for a cylindrical geometry. The growth constant are selected in an arbitrary way. However they are selected in a way where the cooling curves have a shape similar to the one observed experimentally for this type of reaction. The radius of the cylinder is given by a material balance from the number of growing graphite units and the lever rule.

A number of calculations were performed and the results are shown in figures 6 and 7.

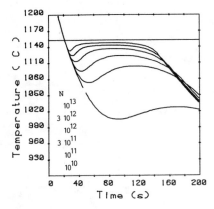
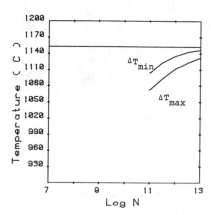

Figure 6.
Cooling curves for vermicular graphite calculated for different numbers of growing nodules.

Figure 7.
Minimum and maximum undercooling as function of cell numbers for vermi-cular graphite.

A comparison between figures 6 and 7 on one hand, figures 2, 3 and 4, 5 on the other hand shows that the undercooling for the growth of vermicular graphite is much higher than for flake graphite but smaller than for nodular graphite. It is also not so sensitive as nodular cast iron to the number of growing units.

According to Hillert (8) the white structures are normally formed in two consecutive steps. One step where primary $Fe_3C$ plates are formed, another where a normal eutectic structure is formed, figure 8. No kinetic law is known for the first growth process. However, based on the model for plate-like precipitation given by Trivedi (9) the growth law given in annex 6 is found. The plates of $Fe_3C$ are growing randomly in the liquid and collide with each other. The impingement or collision between the plates are des-cribed by an Avrami equation given in annex 6.

After the formation of the plates an eutectic structure will be formed. The growth of this structure is described by the equations shown in annex 3. The physical constants used are given in table 1.

By combining the equations in annex 1, 3 and 6, cooling curves were cal-culated. The results are shown in figures 9 and 10.

157

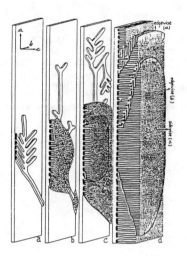

Figure 8.
The growth process of white structure, according to Hillert, Steinhaüser (8).

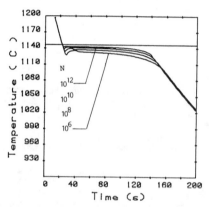

Figure 9.
Cooling curve for white structure for different numbers of growing cementite plates.

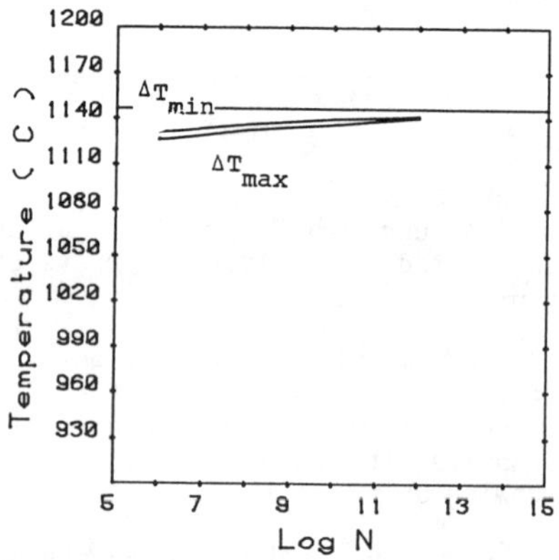

Figure 10.
Minimum and maximum undercooling as function of the number of cementite plates.

A comparison with the earlier presented cooling curves shows that the white structure formed gives the lowest undercooling of all the structures. In relation to the eutectic temperature. The calculations also show that the white structure is not very sensitive with respect to the number of growing units. The explanation for this is the much higher growth rate for the white structure than for another of the structures formed.

### Discussion and concluding remarks

A series of calculations of cooling curves for different types of sub-structures formed in cast iron castings have been performed. It was shown that the shape of the curves are very much influenced by the growth law. This will make it possible to distinguish between different substructures formed directly from a cooling curve. By computer analysing an experiementally found cooling curve the growth constants can be evaluated. When the

158

growth constants are found the type of structure formed can be predicted. This will in the future be a tool for the foundrymen. If the growth constants are known one can also calculate the structure formed in different casting processes. For instance it will be possible to calculate the fraction of white structure in a casting. This is for instance important for thin walled castings.

REFERENCES

1. H. Fredriksson and S.E. Wetterfall, The Metallurgy of Cast Iron (Proceedings of 2nd Int. Symp. on Metallurgy of Cast Iron, Geneve, 29-31 May 1974, Georgi Publishing Comp. St Saphorin, Switzerland.

2. H. Fredriksson and I.L. Svensson, Computer simulation of the structure formed during solidification of cast iron, (3rd Int. Symp. on Metallurgy of Cast Iron 1984, North Holland Publ.)

3. D.M. Stefanescu and C Kanetkar, Computer modelling of the solidification of eutectic alloys: The case of cast iron, (Proceeding Society for Metals Materials Science Division Computer Simulation. The Metallurgical Society in Toronto, Canada, October 13-17, 1985).

4. N. Chworinoff, Die Giesserei, Heft 10 (1940)

5. H. Fredriksson and A. Olsson, On the mechanism of the transition from columnar to equiaxed crystals in ingots, (TRITA-MAC-0177, Aug 1980)

6. M. Hillert and Subba Rao V.V, Grey and White solidification of cast iron (Proceeding of the conference on the solidification of Metals, Brighton, Dec. 4-7, 1967, The Iron Steel Institute).

7. H. Fredriksson, On the mechanism of eutectic growth, (Materials letter, September, 1987).

8. M. Hillert and H. Steinhaüser, (Jernkontorets Annaler 1960), (144), 520.

9. W.P. Bosze and R. Trivedi, (Met. Trans. 1974, (5)), 511.

## Annex 1. Heat transfer from the casting mould

$$\frac{dQ}{dt} = -A_i \sqrt{\frac{K_f \cdot C_p^f \cdot \rho_f}{\pi \cdot t}} \cdot (T_i - T_0) \tag{1}$$

## Annex 2. Precipitation of austenite dendrites

$$\frac{dQ}{dt} = V_0 \cdot \rho_m \cdot \left( C_p^m \cdot \frac{dT}{dt} + \Delta H^a \cdot \frac{df^a}{dt} \cdot N \cdot \frac{4}{3}\pi R^3 + N \cdot f \cdot \Delta H \cdot 4\pi R^2 \frac{dR}{dt} \right) \tag{2}$$

$$\text{until } R_{max} = \sqrt[3]{\frac{V_0}{N} \frac{3}{4\pi}} \tag{3}$$

$$\frac{dR}{dt} = A (\Delta T)^n \tag{4}$$

$$f_s^\gamma = \frac{1 - \dfrac{c_c^0}{c_c^{L/\gamma}}}{1 - k_c^{L/\gamma}} \tag{5}$$

$$c_c^{L/\gamma} = 4.26 + 9.18 \cdot 10^{-3} (1153 - T_L) \tag{6}$$

when $R = R_{max}$ then

$$\frac{dQ}{dt} = V_0 \cdot \rho \cdot (C_p^m \cdot \frac{dT}{dt} + \Delta H \cdot \frac{df}{dt}) \tag{7}$$

constants $A = 2 \cdot 10^{-6}$    $N_g^\gamma = 4 \cdot 10^{-10}$    $N_g^{eut} = 4 \cdot 10^{-10}$

## Annex 3. Eutectic growth

The undercooling as a function of the growth rate and lamellar distance for the two phases are normally divided into three terms, one describing the effect of new surfaces formed, one describing the interface kinetics and one describing the diffusion effect. The model of the growth law follow the procedure suggested by Fredriksson (7). The supersaturation due to the kinetics is in this case neglected.

$$\Delta T = \Delta T_\sigma + \Delta T_{diff}$$

The undercooling due to the formation of new surfaces is normally written:

$$\Delta T_\sigma^\alpha = \frac{2\sigma^{\alpha/\beta} T_E}{f^\alpha H^{eut}_{\lambda \cdot \rho}} \tag{8}$$

$$\Delta T_\sigma^\beta = \frac{2\sigma^{\alpha/\beta} T_E}{f^\beta \Delta H^{eut}_{\lambda \cdot \rho}} \tag{9}$$

Where $\sigma^{\alpha/\beta}$ is the surface tension between the two solid phases $\Delta H^{eut}$ the heat of fusion of the eutectic structure. $\lambda$ the lamellar distance and $T_F$ the melting point of the eutectic.

The supersatoration due to the diffusion ahead of the growing interface will in this case be written:

$$\Delta C_{diff}^\alpha = \frac{V\lambda}{D_s^L} \cdot f^\alpha (C^{\beta/L} - C^{\alpha/L}) \tag{10}$$

$$\Delta C_{diff}^\beta = \frac{V\lambda}{D^L} \cdot f^\beta (C^{\beta/L} - C^{\alpha/L}) \tag{11}$$

By combining equations (5) and (9) for the $\alpha$-phase and equation (6), (8) and (10) for the $\beta$-phase the undercooling for each phase can be evaluated. The undercooling for ezch phase must be equal and one obtains a single expression for the growth rate as a function of the lamellar distance.

$$\Delta T_{diff} = \frac{dT}{dc} \cdot \Delta C_{diff} \tag{12}$$

Annex 4. Growth of Nodular Cast-Iron

The model of solidification process of nodular cast iron prepared by (1) Wetterfall, Fredriksson and Hillert has been used.

$$\frac{dR}{dt} = D_C^\gamma \frac{v_m^{gr}}{v_m^\gamma} \frac{1}{0.243 \cdot R} \frac{(X^{\gamma/L} - X^{\gamma/gr})}{(X^{gr} - X^{\gamma/gr})} \tag{13}$$

$$D_C^\gamma = 9 \cdot 10^{-11} \text{ m}^2$$

$$v_m^{gr} = 5.5 \cdot 10^{-6} \text{ (m}^3/\text{mole)} \qquad X^{\gamma/L} - X^{\gamma/gr} = 3.66 \cdot 10^{-4} \cdot \Delta T$$

$$v_m^\gamma = 7.0 \cdot 10^{-6} \text{ (m}^3/\text{mole)} \qquad X^{gr} - X^{\gamma/gr} = 0.909$$

The solidification rate

$$\frac{df}{dt} = (1 - fs) \, 4 \, \pi R^2 \cdot \frac{dR}{dt} \cdot N \tag{14}$$

161

## Annex 5. Vermicular graphite

The growth of vermicular graphite has been modelled by using two different growth law on a cylindrical shape.

The growth of radius will be through diffusion of carbon through the solid austenite like growth of nodular cast-iron (see annex 4).

The other growth direction, a linear law is applied

$$\frac{dy}{dt} = 2 \cdot 10^{-9} \cdot \Delta T \tag{15}$$

The solidification rate will be

$$\frac{df}{dt} = (1 - fs) \ (2\pi R \cdot Y\frac{dR}{dt} \cdot N + \pi R^Z \cdot \frac{dy}{dt} \cdot N) \tag{16}$$

## Annex 6. White structure

The white eutecticum is modelled by assuming that a plate of $Fe_3C$ and autenite is growing in four directions

$$\frac{dy}{dt} = 4 \cdot 10^{-10} \ (\Delta T)^4 \tag{17}$$

$$\frac{df}{dt} = (1-fs) \ 4 \cdot Y \cdot Z \cdot \frac{dy}{dt} \cdot N \tag{18}$$

The coarsness, Z, of the eutectic plate is given of the phasediagram and number of growing plates, at the growth temperature. When the plates has reached the maximum size, an eutectic growth law (6) controls the growth rate, perpendicular to the plate. The eutectic white-reaction has been calculated by the models shown in annex 3.

$$\frac{df}{dt} = (1 - fs) \ Y^2 \cdot N \cdot \frac{dZ}{dt} \tag{19}$$

# PREDICTING RESIDUAL STRESSES IN IRON CASTINGS

J. W. Wiese and J. A. Dantzig

University of Illinois at Urbana-Champaign
Department of Mechanical and Industrial Engineering
144 Mechanical Engineering Building
1206 West Green Street
Urbana, IL 61820

## Abstract

A method is being developed to predict residual stresses in cast iron foundry castings. Using a commercial finite element package with a user-written stress element formulation, stresses resulting from thermal displacements in the cooling casting are computed, accounting for cast iron's differing mechanical behavior in tension and compression. A user-written element is also provided to apply force-displacement boundary conditions to the surface of the casting, eliminating the mold's finite element mesh from the problem. The mold was previously eliminated from the heat transfer calculation through the application of the boundary curvature method, which provides CPU time savings of up to 95 percent while maintaining accuracy in the three dimensional solidification solution. This new method for stress, along with the boundary curvature method for the thermal analysis, allows for the correct solution of solidification-stress problems in cast iron in only a small fraction of the time previously necessary.

Solidification Processing of Eutectic Alloys
D.M. Stefanescu, G.J. Abbaschian and R.J. Bayuzick
The Metallurgical Society, 1988

## Introduction

Foundry casting is a popular fabrication method because of its adaptability to complex geometrical shapes and relatively low cost. The most expensive aspects of developing a new casting are the numerous prototypes needed for evaluation and the labor needed to produce them. A trial and error process must be followed until a new design is free of defects, such as porosity, and high levels of residual stress. Casting stresses can lead to failure while still in the mold. The goal of this research is to reduce the cost of prototyping by providing the designer with a set of simulation tools to discover problem areas and test solutions before committing to production.

The finite element method (FEM) was chosen because of its ability to model arbitrary geometries, and to incorporate complex temperature dependent material behavior in both the thermal and mechanical properties. The software is integrated into existing FEM codes, the one used most often being ANSYS. The software is intended to be as easy to use as possible, to reduce the computer run time compared to codes presently available, and most important, to obtain correct solutions to general three dimensional problems.

Cast iron is a somewhat unusual engineering material due to its microstructure. Figure 1 shows a typical microstructure of a gray cast iron, illustrating that the material consists of a steel matrix containing thin flakes of graphite. These flakes bear no load when the iron is under tensile stress, resulting in a bulk material with radically different properties in tension and compression [1]. Under tension the flakes open up and initiate cracking, while under compression the flakes close. Figure 2 compares the tension and compression properties of a typical cast iron. The ratio of strength in compression to that in tension can be anywhere from two to four. Present FEM codes have not dealt with this asymmetry of properties over the temperature ranges associated with solidification processes, so new methods are needed.

After incorporating these new tools, there are five steps in the analysis of a casting design:

1. Geometric model and FEM mesh.

2. Determination of material properties.

3. Thermal solution.

4. Stress solution.

5. Evaluation and redesign.

Each step is discussed below.

## Thermal Solution

It is important to obtain the proper thermal solution because there is an intimate connection between the thermal history, the microstructure, and the properties of the cast iron. While we do not model the microstructure development explicitly, we do include thermal and mechanical properties, which vary strongly with temperature, and these, along with the boundary conditions, determine the final stress state of the casting.

164

Figure 1: Micrograph of a typical gray cast iron.

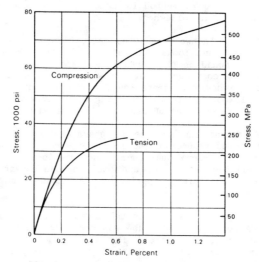

Figure 2: Comparison of the stress-strain response for typical gray cast
iron in tension and compression.

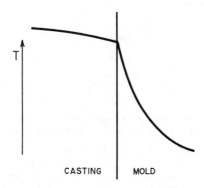

Figure 3: Relative temperature distributions in the casting (cast iron) and
the mold (sand) during the solidification process.

A difficulty in modeling the thermal history of castings can be seen by examining the relative temperature distributions in the iron casting and the sand mold (Fig. 3). Because the thermal diffusivity is so much higher in the iron than in the sand, the thermal gradients in the casting are fairly small, while those in the mold are large. To properly represent high gradients in the FEM, a fine mesh must be used, resulting in a model with most of the nodes in the mold. However, the designer is generally not interested in the details of the thermal field in the mold. A method for eliminating the mold from the problem has been found using a technique called the boundary curvature method [2].

The boundary curvature method consists of representing an area on the surface of the casting as a simple shape such as a plane, cylinder, or sphere, and applying a heat flux boundary condition based on that shape. The representation is based on local topography, where "local" is defined by a length scale $\ell = (\alpha t)^{1/2}$, where $\alpha$ is the thermal diffusivity of the mold material, and t is time. The method is automated in a program called SPIDER which is described in detail elsewhere [3,4].

As an illustration in two dimensions (Fig. 4), the method starts at the centroid of an element and looks out in all directions a distance $\ell$. Two intercept points are found on the surface, and these points, along with the centroid, are used to construct a circle with radius R. This circle, or cylinder, becomes the new representation for the element. It is important to note that the element itself is not changed in any way, rather, the heat flux corresponding to a cylinder of radius R is applied to its surface.

In three dimensions the surface is characterized by two curvatures, called the principal curvatures. The possible surface representations now include spheres (two non-zero curvatures) in addition to planes (two zero curvatures) and cylinders (one zero and one non-zero curvature). Figure 5 shows the boundary conditions applied to a complex three dimensional casting.

The thermal analysis of a cube is presented to demonstrate the benefits of the method. One eighth of the cube was modeled, and the simulation was run both with the mold mesh and using the boundary curvature method. Successive positons of the solidus are shown in Figure 6, demonstrating that the solutions were equivalent using both the new and old methods. The boundary curvature method does well, with less than 2% error in the solidification time. The solidification time is tracked because it is very sensitive to errors in latent heat evolution.

Table 1

| Casting Elements | Mold Elements | Boundary Condition Elements | ANSYS Run Time* | SPIDER Run Time* |
|---|---|---|---|---|
| 64 | 576 | -- | 23730 | --- |
| 64 | --- | 96 | 1883 | 5.5 |

* CPU times in seconds on a Ridge 32C supermini computer

Run times are compared in Table 1. Note that in the first case 90% of the elements are in the mold, and that the run utilizing SPIDER was able to save over 92% of the run time compared to the one including the mold.

166

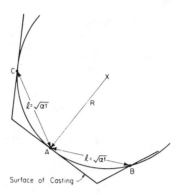

Figure 4: Two dimensional illustration of the boundary curvature method.

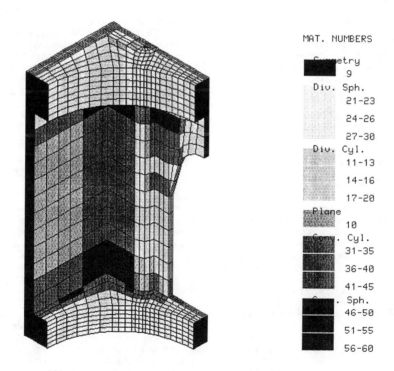

Figure 5: Boundary conditions as applied by SPIDER to a section of an engine block casting.

The accuracy of the solution, no matter what method is chosen, is greatly affected by the choice of time step and element sizes. Figure 7 shows the error in computed solidification time versus time step size, for various numbers of elements through the half thickness. This figure is valid for cast iron, using an effective specific heat for the latent heat of fusion, and shows that the error increases for increasing time step size and decreasing number of elements. The best choice for cast iron is 200 steps through the solidification time of the smallest section of interest, and four elements through the half thickness.

## Stress Solution

In performing the stress solution, the benefits of excluding the mold from the problem should be maintained, requiring development of a new mechanical boundary condition element. This was based on an existing surface effect element, STIF52 in ANSYS [5]. In an actual casting situation, the sand crushes and does not spring back to its former shape. In contrast, if the casting moves away from the mold, no resistance is offered. Modeling this behavior can be done with an element having the three components shown in Figure 8; a spring to offer resistance when the mold is crushed, a gap which opens if the casting moves away from the mold, and a ratchet which prevents the sand from rebounding elastically. Together, these components should accurately represent the behavior of the sand mold. The boundary condition element is still under development.

Perhaps the most important aspect of predicting residual stresses in these simulations is the proper characterization of cast iron's asymmetric yielding behavior. The behavior has been added to an existing eight noded isoparametric element, STIF45 [5], by giving it the ability to model a complex yield surface (Fig. 9). The yield surface is based on a failure surface described by Frishmuth and McLaughlin in 1976 [6], and retains the general shape of that surface. The shape of the new yield surface is geometrically similar to the shape of Frishmuth and McLaughlin's failure surface, obtained by multiplying the failure stresses by .25. This reflects the common practice of assuming a yield that is proportional to the ultimate strength in cast irons, the most common factor being 25% [1]. The equations used by Frishmuth and McLaughlin are not convenient for fitting the yield surface to experimental data, such as the yield stresses in tension and compression, so other mathematical forms are used. These forms are described below.

There are four domains on the new yield surface: (1) Three tensile principal stresses (TTT). In this region the flakes open up, and all three principal directions may act independently. Therefore, the yield surface consists of three mutually perpendicular planes (a cube). (2) Three compressive principal stresses (CCC). The flakes are compressed and closed, resulting in behavior similar to that of solid steel, so a Von Mises yield surface is used. This is simply a circular cylinder inclined equally to the three principal directions in principal stress space. (3) One tensile and two compressive stresses (TCC). In this area a surface is needed to smoothly join the relatively low yield in tension to the high yield in compression. A paraboloid of revolution was chosen to represent this domain. (4) Two tensile and one compressive stress (TCC). A blending function is used to smoothly merge the cube of the TTT region with the parabola of the TCC region. Figure 10 illustrates the match between the new yield surface and experimental measurements presented by Coffin [7].

The yield surface is used during the stress solution to construct stress-strain curves. Because these curves are nonlinear and temperature

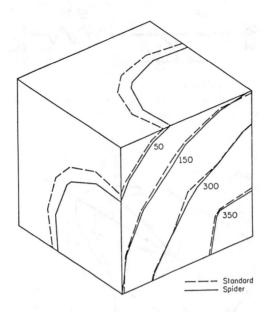

Figure 6: Progressive solidus locations in one-eighth section of a 60 mm edge cast iron cube. The lower right front face, the right rear face, and the bottom rear face are planes of symmetry. Times are in seconds from pouring.

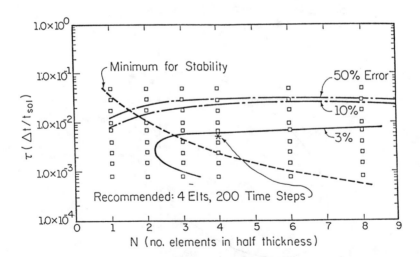

Figure 7: Plot of solidification time errors (absolute values) versus time step size and number of elements through the half thickness for a series of 1-D simulations. The dimensionless time step size measured along the ordinate is the actual time step divided by an estimation of the solidification time.

169

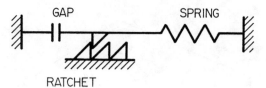

Figure 8: Schematic representation of the proposed mechanical boundary condition element.

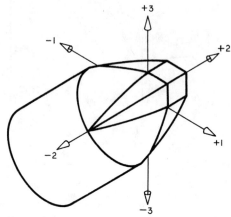

Figure 9: Illustration of the new cast iron yield surface in 3-D principal stress space.

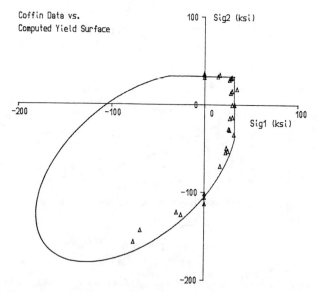

Figure 10: Comparison of new yield surface to experimental data from reference 7. Data are from biaxial tests, where Sig1 and Sig2 represent two principal stresses.

170

dependent, an iterative solution is required at each stress solution step. ANSYS works by saving the temperatures for each thermal time step in a file [8]. These temperatures are later read by the stress analysis and applied to the model. The thermal and stress analyses are decoupled because the heat of deformation is negligible, and the shrinkage is assumed to have no effect onthe thermal boundary conditions.

With our alterations, during each iteration of the stress solution for a particular element, the principal stresses from the last iteration are used to form a vector in stress space. The point at which that vector pierces the yield surface is utilized to compute a yield stress and stress-strain curve. Note that the yield surface consist of the several pieces described above. Thus, the material behavior depends not only on the temperature, as it does in the unmodified element, but also on the stress state.

Example

Figure 11 shows the temperature solution for a tapered iron slab at 550 seconds after pouring. The temperatures are just above the eutectoid temperature for this material. Figure 12 shows the stresses in the vertical direction computed using symmetric tension and compression properties and the unaltered stress element. The properties for tension were reflected into the compression region resulting in a simple Von Mises yield surface. For the results shown in Figure 13, the new yield surface was used with a three to one ratio of compression to tension properties. Notice that for the first case the maximum and minimum stresses are 0.93 and -1.17 MPa, respectively, while for the second they are 1.13 and -2.77 MPa. When continued to room temperature, the result would be the residual stresses.

Discussion

For the second instance above, using the new yield surface, the maximum tensile stress is 24% higher, while the maximum (absolute value) compressive stress is 137% higher. Since the yield points in compression are higher, three times in this case, compressive stresses do not have as great an opportunity to relax through plastic strain. Similarly, the higher yield points result in smaller compressive strains, causing the areas under tension to be stressed to higher levels.

As in the thermal solution, the issue arises as to the proper number of steps to take through the simulation, and the number of elements to use. Generally, the number of elements required for a proper thermal solution is more than enough for the stress solution, but time step size is a different matter. Figure 14a illustrates this point for a simplified residual stress calculation.

At an early time and high temperature, the mechanical properties are represented by the bottom curve. Thermal stress is driven by strains, so for the first step the strain is $e_1$. The intersection with the stress-strain curve is found at point A, and the plastic strain $e_{p1}$ is used as the offset, assuming kinematic hardening, for the next step. The process is repeated, accumulating plastic strain at each step, until the final state is reached with a residual stress of $S_{R1}$. Figure 14b shows a similar process, but in this case only two steps are taken. By comparing Figure 14a to 14b, it can be seen that the residual stress in the second case, $S_{R2}$, is higher than $S_{R1}$ due to a smaller accumulation of plastic strain. The second process yields estimations for the residual stress that are too high. Preliminary results show that one stress step for every three thermal solution step yields consistent results.

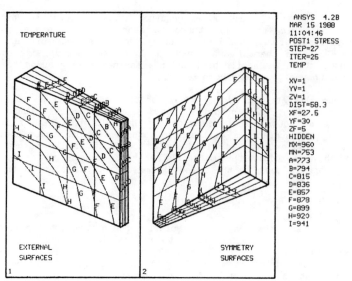

Figure 11: Temperature solution for tapered iron slab example problem. One eighth of the object is modeled. Dimensions along the long sides are 60 mm, the center is 15 mm thick, and the edge is 5 mm thick. The view to the left shows the external planes and the one to the right shows the symmetry planes. Temperatures are in degrees Celcius.

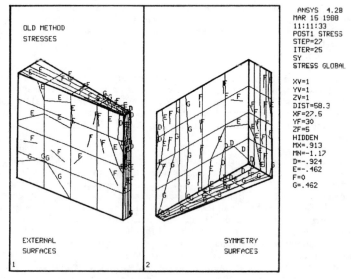

Figure 12: Stresses in the y (vertical) direction for the model of fig. 11. The Von Mises yield function was used, units are MPa.

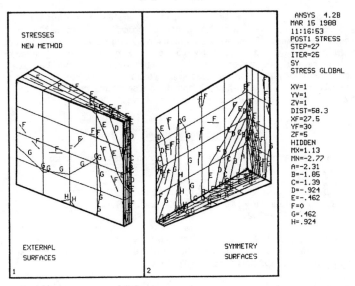

ANSYS 4.2B
MAR 15 1988
11:16:53
POST1 STRESS
STEP=27
ITER=25
SY
STRESS GLOBAL

XV=1
YV=1
ZV=1
DIST=58.3
XF=27.5
YF=30
ZF=5
HIDDEN
MX=1.13
MN=-2.77
A=-2.31
B=-1.85
C=-1.39
D=-.924
E=-.462
F=0
G=.462
H=.924

STRESSES
NEW METHOD

EXTERNAL
SURFACES

SYMMETRY
SURFACES

Figure 13: Stresses in the y (vertical) direction for the model of fig. 11. The new yield surface was used, units are MPa.

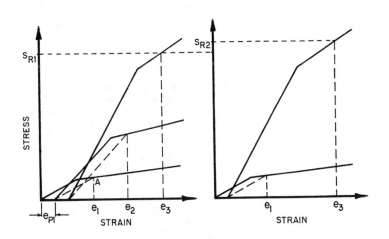

Figure 14: Graphical representation of thermal stress calculations. Lower curves are at higher temperatures. a) Three steps taken during solidification. b) Two steps taken during solidification.

173

## Conclusions

A set of tools is being developed to enable the engineer to evaluate casting designs on the computer before committing valuable resources to prototype development. These tools consist of software that interfaces with existing FEM packages and is easy to use. For the thermal analysis, using SPIDER can reduce computer run times by over 90% while yielding well documented accuracy. For stress analysis, new element formulations are under development which show great promise for run time reduction and accuracy in representing stresses in cast iron.

## References

1. C. F. Walton and T. J. Opar, eds., Iron Castings Handbook (Iron Casting Society, Inc., 1981).

2. J. A. Dantzig and S. C. Lu, "Modeling of Heat Flow in Sand Castings: Part I. The Boundary Curvature Method," Metallurgical Transactions B, 16B(2)(1985), 195-202.

3. J. A. Dantzig and J. W. Wiese, "Modeling of Heat Flow in Sand Castings: Part II. Applications of the Boundary Curvature Method," Metallurgical Transactions B, 16B(2)(1985), 203-209.

4. J. A. Dantzig and J. W. Wiese, "Modeling the Solidification of Foundry Castings," Advanced Manufacturing Processes, 1(3&4), 437-454.

5. P. C. Kohnke, ANSYS Engineering Analysis System Theoretical Manual (Houston, PA: Swanson Analysis Systems, Inc., 1983).

6. R. E. Frishmuth and P. V. McLaughlin, "Failure Analysis of Cast Irons Under General Three-Dimensional Stress States," Journal of Engineering Materials and Technology, 69(1)(1976), 69-75.

7. L. F. Coffin, Jr., "The Flow and Fracture of a Brittle Material," Journal of Applied Mechanics, 17(3)(1950), 233-248.

8. Gabriel J. DeSalvo and John A. Swanson, ANSYS Engineering Analysis System User's Manual Volume I, Houston, PA: Swanson Analysis Systems, Inc., 1985).

174

# UNDERCOOLED AND
# RAPIDLY SOLIDIFIED ALLOYS

SUPERCOOLING EFFECTS IN FACETED EUTECTIC Nb-Si ALLOYS

A.B. Gokhale, G. Sarkar, G.J. Abbaschian,
J.C. Haygarth*, C. Wojcik*, and R.E. Lewis**

University of Florida, Gainesville, FL 32611
* Teledyne Wah Chang Albany, Albany, ORE 97321
** Lockheed Missiles and Space Company, Inc., Palo Alto, CA 94304

Abstract

The eutectic alloy Nb-58 at.% Si was melted and supercooled in an electromagnetic (EM) levitation apparatus. The changes in the microstructural morphology and compositions of the intermediate phases $Nb_5Si_3$ ($\gamma$) and $NbSi_2$ ($\delta$) in the solidified samples were studied as a function of the degree of melt supercooling, holding time in the supercooled state, and the cooling rate. It was found that the morphology, distribution, as well as the volume fractions of the phases, are related to the nucleation sequence and growth behavior of the faceted intermediate phases. The primary phase nucleation in the alloy occurred usually at a bulk supercooling of approximately 60 K. The primary phase was found to be the faceted intermediate phase $\gamma$. Upon further cooling, the remaining liquid continued to supercool by an appreciable extent, even in the presence of solid particles, until the second phase, $\delta$, nucleated. The growth of $\delta$ was accompanied by a rapid recalescence. The final liquid to solidify formed a coupled eutectic with Nb content greater than the original eutectic composition. When the primary phase nucleation was suppressed, a fine cellular eutectic with quasi-regular morphology was obtained. The results are discussed based on various aspects of eutectic growth from a supercooled melt. An evaluation of Nb-Si phase diagram is also presented.

Solidification Processing of Eutectic Alloys
D.M. Stefanescu, G.J. Abbaschian and R.J. Bayuzick
The Metallurgical Society, 1988

# Introduction

Nb-based alloys are currently considered as possible candidates in high-temperature oxidation-resistant structural applications. The high temperature oxidation resistance of Nb-based alloys can be improved by coating either with pure Si[1] or silicides like MoSi$_2$.[2,3] However, these methods of improving the oxidation resistance suffer from the disadvantage that a breakdown in the integrity of the coating can result in accelerated oxidation, leading to a serious deterioration of the mechanical properties. As an alternative, it has been hypothesized that alloying with Si may yield a self-healing surface film of SiO$_2$ through a continuous supply of silicon atoms to the surface.

Alloying with Si also substantially increases the hardness and yield strength but is accompanied by a considerable loss in ductility. Three intermediate phases (Nb$_3$Si, Nb$_5$Si$_3$, and NbSi$_2$) are formed in the Nb-Si system. Each of these is brittle, the hardest and most brittle being Nb$_5$Si$_3$. Thus, the practical use of Nb-Si alloys is limited by their brittleness. Consequently, the possible enhancement of ductility in these alloys through various processing routes (e.g., solidification processing or mechanical alloying) is of great technical importance.

Any attempts at improving the ductility must consider the crack initiation and propagation behavior, both of which depend, among other factors, on the scale and morphology of the microstructure, and the spatial distribution of the constituent phases. As part of our current research, we have processed Nb-Si alloys by melt supercooling and rapid solidification with the aim of producing refined microstructures, which may ultimately lead to improved mechanical properties of Nb-Si alloys.

Morphological analysis of the solidification microstructures of eutectic and off-eutectic alloys is generally based on the shape and extent of the coupled zone, which delineates the temperature-composition boundaries of the region of stability for coupled eutectic growth. The origin of the coupled growth zone lies in the relative change of the growth (interface) temperature of various morphological forms as a function of growth velocity and composition. It is generally accepted[4] that, under a given set of growth conditions, the growth form with the highest interface temperature, i.e., lowest interfacial undercooling, will be more stable, and will "lead" the growth front.

The shape of the coupled zone depends upon the relative growth difficulties of the constituent phases. When both phases are non-facet forming, the coupled zone is symmetrical with respect to the eutectic point. For a faceted/non-faceted eutectic, on the other hand, the facet forming constituent will experience kinetic attachment difficulties during growth, resulting in a skew in the coupled zone towards the faceted phase, as illustrated by the shaded region in Figure 1(a). In this example, the α phase is shown to be facet forming, while β is non-faceted. For the alloy C$_0$ in Figure 1(a), the interface temperatures for the various growth forms as a function of growth velocity are shown schematically in Figure 1(b). In this figure, "E" indicates coupled eutectic front. At any given growth velocity, the most stable form is given by the solid line. The corresponding shape of the coupled zone is shown in Figure 1(a) by the shaded region. Thus, for the alloy C$_0$, for small growth velocities, coupled eutectic front will be more stable (points a to b). At higher growth velocities and therefore larger interfacial supercoolings (points b to c), the α phase will lead the growth front, while between the points c and d, the coupled eutectic front will again be the most stable growth form.  At still higher growth velocities, β dendrites will lead the growth front.

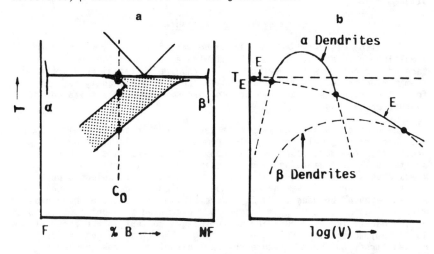

Figure 1:  Schematic representation of the coupled growth zone in faceted/ non-faceted eutectics.  (a) Shape of the zone.  (b) Interface temperature of various growth forms as a function of growth velocity.

The implicit underlying assumption in the above reasoning is that both solid phases are already present so that the growth of the phase (or phases) can take place accordingly.  This is usually the case during directional solidification.  Indeed, the technique has been used to experimentally determine the coupled zone.  By contrast, the above mentioned assumption may not be true during solidification processing through bulk supercooling.  Under these conditions, even inside the coupled growth zone, one of the solids may nucleate first and lead the growth, depending upon the relative nucleation probability of the constituents.  As a result, the composition of the remaining liquid will shift and possibly move out of the coupled zone to an extent where the growth of the other phase (if and when it nucleates) becomes preferred.

This behavior is brought out clearly by our experiments with Nb-Si eutectic alloys, where it is shown that starting with an alloy of eutectic composition, various growth morphologies can be obtained depending on the nucleation sequence.

This paper describes the effect of melt supercooling on the microstructure of the Nb-58 at.% Si alloy.  The eutectic alloy (between the intermediate phases $Nb_5Si_3$ and $NbSi_2$) was processed by electromagnetic levitation melting, followed by rapid quenching from the supercooled state.  It was found that starting with an alloy nominally of eutectic composition, nucleation of $Nb_5Si_3$ occurred in the supercooled liquid first.  Upon further cooling, the remaining liquid continued to supercool to a larger extent until the second phase, $NbSi_2$, nucleated. The latter was commonly accompanied by a rapid recalescence, and the remaining liquid finally entered the coupled growth zone.  The primary phase itself exhibited several interesting features including a eutectoid-type decomposition.  These and other observations are discussed on the basis of quantitative microstructural measurements, compositional and thermal analysis, and preliminary thermodynamic modeling of the phase diagram.

## Experimental Procedure

The alloy was received from Teledyne Wah Chang Albany in the form of a multiple-pass arc melted button weighing approximately 50 gm. Once face of the button was prepared metallographically for microstructural examination as well as microhardness and compositional analysis, following which the button was broken into smaller pieces suitable for further processing.

The processing and analysis were carried out in the following sequence: (1) arc melting, (2) electromagnetic (EM) levitation processing, (3) metallography, (4) compositional analysis and back scattered electron imaging, and (5) quantitative microstructural measurements. Arc melting was carried out on a water cooled copper hearth in a partially evacuated chamber in a purified argon atmosphere. Pieces weighing between 1 and 2 gm were placed on the copper hearth along with a titanium getter. The chamber was evacuated to a pressure of 50 μm Hg and backfilled with purified argon (oxygen content on the order of $10^{-11}$ ppm) three times before arc melting. The titanium getter was melted first, to further reduce the oxygen content in the chamber, followed by arc melting of the samples.

The arc melted samples were levitated in an electromagnetic levitation coil using a 15 kW RF generator. The details of the EM levitator have been given previously by Amaya, et.al.[5] A Pyrex glass tube of 22 mm outer diameter was placed inside the coil, through which a mixture of purified Ar and He was passed to maintain an inert atmosphere around the sample during levitation. The sample temperature, which could reach as high as 3000 K, was monitored continuously by a single color optical pyrometer connected to a strip-chart recorder. The pyrometer was calibrated for sample emissivity using pure niobium as a standard. The accuracy of temperature measurement was ± 10 °C. After processing, the samples were quenched on a copper plate or in a moving piston type splat cooler.

The processed samples were prepared metallographically using standard techniques on a section which allowed for examination of the microstructure from the surface in contact with copper quenching plate to the one away from it. Back scattered electron imaging of the as-polished samples was used for phase differentiation based on average atomic number contrast. A JEOL 733 electron microprobe and pure Nb and Si standards were used for the compositional analysis. Eutectic spacing measurements were done on the electron microprobe by digital image processing using a DEC RT-11 computer and Tracor-Northern software. After the analysis, the samples were etched and volume fractions of various phases and/or phase mixtures were measured optically by standard stereological techniques.

## Results and Discussion

### Nb-Si Phase Diagram

Since a recent evaluation of the phase diagram is not available, and the commonly used diagram from Mofatt[6] or Metals Handbook[7] contain errors or omissions, the diagram was evaluated in the present investigation. The actual diagram is based on the more recent work of Kocherzhinskii, et.al.[8] The system exhibits three intermediate phases, $Nb_3Si$, $Nb_5Si_3$, and $NbSi_2$. We have adopted the following nomenclature for the various phases: (Nb) ≡ α, (Si) ≡ β, $Nb_3Si$ ≡ ε, $\alpha Nb_5Si_3$ (low temperature) ≡ $\gamma_1$, $\beta Nb_5Si_3$ (high temperature) ≡ $\gamma_2$, and $NbSi_2$ ≡ δ. $Nb_3Si$ forms peritectically at 1980 °C while $Nb_5Si_3$ and $NbSi_2$ melt congruently at 2520 and 1940 °C, respectively. The diagram shows three eutectics: between (Nb) and $Nb_3Si$ at 1920 °C and 17.5 at.% Si, between $Nb_5Si_3$ and $NbSi_2$ at 1900 °C and 58 at.% Si, and between $NbSi_2$ and (Si) at 1400 °C and 98 at.% Si. The intermediate phase $Nb_3Si$ decomposes eutectoidally at 1770 °C, while the intermediate phase $Nb_5Si_3$ transforms to a low temperature form in the range 1940 to 1650°C. According to (8), the latter transformation involves a eutectoid-type decomposition, $\beta Nb_5Si_3 = \alpha Nb_5Si_3 + NbSi_2$.(9) A similar low temperature transformation was also reported by Samsonov(9) and Pan et.al.(10) The terminal solid solution (Nb) has a maximum solubility of 1.2 at.% Si at 1920 °C. $Nb_3Si$ and $NbSi_2$ are essentially stoichiometric line compounds, while $Nb_5Si_3$ exhibits a homogeneity range of 37 to 40.5 at.% Si. As discussed below, our microprobe data showed excellent agreement with these compositions, except for the solid solubility of (Nb), which we found to be close to 2 at.% Si.

$Nb_3Si$ is tetragonal $Ti_3P$-type with lattice spacings of a = 1.023 and c = 0.5189 nm.(10,11) $\alpha Nb_5Si_3$ is tetragonal $Cr_5B_3$-type with lattice spacings of a = 0.6571 and c = 1.1889 nm, while $\beta Nb_5Si_3$ is tetragonal $W_5Si_3$-type with lattice spacings of a = 1.0040 and c = 0.5081 nm. $NbSi_2$ is hexagonal $CrSi_2$-type with lattice spacings of a = 0.4791 and c = 0.6588 nm.(8) Density calculations on the basis of lattice parameters and crystal structures gave the following results; $Nb_3Si$ - 7.52, $\alpha Nb_5Si_3$ - 7.104, $\beta Nb_5Si_3$ - 8.38, and $NbSi_2$ - 5.65 g/$cm^3$.

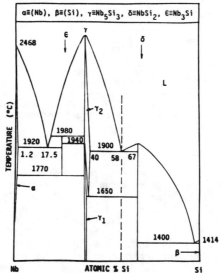

Figure 2: The Nb-Si phase diagram, based on the data of Kocherzhinskii(8). The processed alloy (Nb-58 at.% Si) is shown by the dashed line.

No thermodynamic investigation of the system has been carried out, except in a very limited composition range (0.9 ≤ $X_{Si}$ ≤ 1) by (1). Their findings indicated the liquid solution to be exothermic. In the present investigation, preliminary thermodynamic modeling of the phase equilibria

181

was carried out using the experimental liquidus data and the Gibbs free
energy of fusion of the pure components.

Our calculations also indicate the mixing process in the liquid to be
exothermic. The enthalphy of mixing in the liquid was represented by,

$$\Delta_{mix}H = X_{Si}(1-X_{Si})(-103.9 + 42.62\ X_{Si} - 93.7\ X_{Si}^2)\ kJ/mol$$

where $X_{Si}$ is the mole fraction of Si. Using the heat capacity and heat
of fusion data for Nb and Si taken from (12), the Gibbs energy of fusion
for the pure components was expressed as a function of temperature as:

$$\Delta_{fus}G_{Nb} = 14.73 \times 10^3 + 66.29T + 20.08 \times 10^{-4}\ T^2 - 9.75\ T\ln T \qquad J/mol$$

$$\Delta_{fus}G_{Si} = 48.53 \times 10^3 + 0.497T + 0.193 \times 10^{-2}\ T^2 - 4.37\ T\ln T$$

$$-176.98 \times 10^3 T^{-1} \qquad J/mol$$

where T is in K. The entropy of mixing was assumed to be ideal
(subregular solution model). Based on these expressions, a phase diagram
was calculated.[13] The calculated diagram agreed well with the
experimentally determined equilibria and further predicted that the high
temperature ($\beta$) form of $Nb_5Si_3$ was unstable below 1650 °C, in excellent
agreement with the data of (8). The calculated Gibbs energies of fusion
of the intermediate phases are as follows:

$$\Delta_{fus}G_{Nb_3Si} = 73672 - 32.6\ T \qquad J/mol$$

$$\Delta_{fus}G_{Nb_5Si_3} = 25816 - 9.3\ T \qquad J/mol$$

$$\Delta_{fus}G_{NbSi_2} = 96332 - 44\ T \qquad J/mol$$

It may be noted that the calculated heat of fusion for $NbSi_2$ is
approximately 3.5 times higher than that for $Nb_5Si_3$. Using parameters of
the thermodynamic model, metastable extensions for the (L + S) were
calculated. The $T_o$ curves ($G^L = G^S$) for these phases are expected to be
very close to the solidus boundary of the intermediate phases or its
metastable extension.

## Microstructural Analysis

During EM levitation processing, the samples showed interesting
thermal characteristics upon supercooling, as schematically indicated
with typical time-temperature outputs in Figure 3. The various regions
of interest indicated in Figure 3(a) are as follows: (i) sample heat-up,
(ii) liquidus arrest at the eutectic temperature, (iii) superheating
following complete melting, (iv) sample cooling after attainment of
desired superheat, (v) liquid supercooling and nucleation of the primary
phase at $T_{n1}$, (vi) further supercooling at a reduced cooling rate,
indicating the release of heat of fusion due to continued primary phase
nucleation and growth, (vii) second phase nucleation and massive
recalescence at $T_{n2}$, raising the temperature nearly up to the eutectic
arrest, and (viii) further cooling during levitation.

During supercooling, a majority of the samples exhibited a primary
phase nucleation which was later found to be the intermediate phase
$Nb_5Si_3$. The primary phase nucleation usually took place around 60 K
supercooling and was accompanied by a small thermal arrest. Small solid

182

particles on the surface of the liquid droplet were also visually observed at this time. Secondary nucleation was observed generally with additional supercooling of about 100 K below the primary nucleation. Two other typical thermal histories are shown in Figures 3(b) and 3(c). Figure 3(b) shows the thermal history of a sample in which the primary phase nucleation was delayed until a relatively high supercooling was reached. In this case, the primary phase nucleation at $T_{p1}$ is accompanied by an appreciable recalescence, after which, further supercooling follows the behavior described in Figure 3(a). When primary phase nucleation did not take place, the sample could be held in the supercooled state for long periods of time before recalescence, as shown in Figure 3(c).

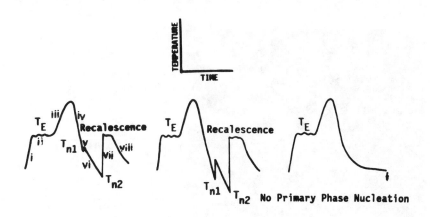

Figure 3:    Thermal effects during processing of three typical samples: (a) A sample which exhibited primary phase nucleation at $T_{n1}$ during supercooling. Second phase nucleation occurred at $T_{n2}$, accompanied by a massive recalescence. (b) A sample in which the primary phase nucleation was delayed to a relatively higher supercooling. (c) A sample in which primary phase nucleation was suppressed and the supercooled state was maintained for an extended period of time.

We were able to quench the samples at any time during the processing cycles shown in Figure 3. As a result, a wide variety of microstructures were obtained as described below. Four typical microstructural morphologies are shown in Figures 4 to 7, together with their experimental cooling curves. The temperature indicated by the arrow is the one at which the sample was dropped from the levitation coil. The sample cools further during free fall until it hits the copper chill plate. Heat transfer calculations indicate that an additional cooling of about 25 K can be expected during free fall.

Figure 4(a) shows the thermal history of a sample which was quenched from a superheated state. The microstructure is predominantly eutectic with a small volume fraction (<1%) of $\gamma$ platelets, as shown in Figure 4(b). This indicates that the alloy is either of a slightly hypoeutectic composition, or upon quenching, $\gamma$ phase nucleated first. A magnified view of the $\gamma$ plates is shown in Figure 4(c). The plate (or flake) shaped $\gamma$ contained 40 at.% Si, while a large beam analysis of the matrix composition indicated an average composition of 58 at.% Si. Note that the $\gamma$ plate is faceted and contains numerous cracks along its width.

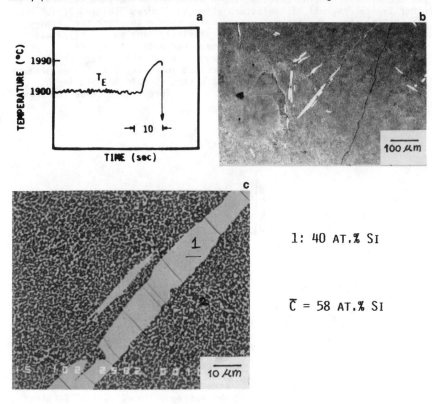

1: 40 AT.% SI

$\bar{C}$ = 58 AT.% SI

Figure 4: (a) Thermal history of a sample quenched from a superheating of approximately 90 K. (b) General microstructural appearance. (c) Detail of $\gamma$ plates (or flakes) in an irregular eutectic matrix.

Figure 5(a) shows the thermal history of a sample quenched immediately following the primary phase nucleation indicated by $T_{n1}$. The general appearance of the microstructure is shown in Figure 5(b). The microstructure is again predominantly eutectic but with $\gamma$ plates coarser than those found in the superheated sample. Figure 5(c) shows the structure near a branched $\gamma$ plate. The $\gamma$ plate contains 41 at.% Si, while the $\delta$ (NbSi$_2$) near the $\gamma$ plate contains 65 at.% Si. The eutectic matrix is shown in Figure 5(d). The irregular eutectic has an average composition of 56 at.% Si, with the compositions of the constituents ($\gamma$ and $\delta$) being 41 and 66 at.% Si, respectively. The compositional analysis indicates a slight metastable extension in the solid solubility of $\gamma$ and $\delta$ at this supercooling.

The thermal history of a sample supercooled to 125 K without primary phase nucleation and held in the supercooled state for 90 s before quenching is shown in Figure 6(a). The microstructure is made up entirely of fine cellular coupled eutectic. Figure 6(b) and 6(c) show

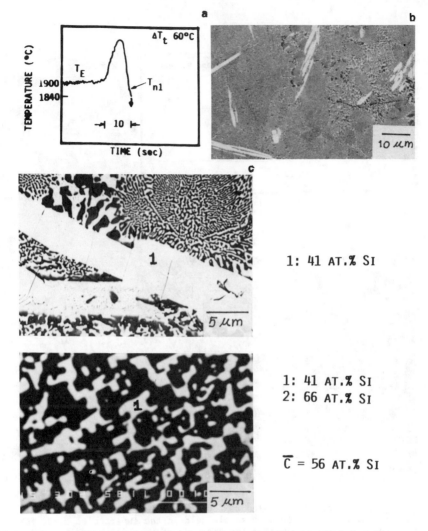

1: 41 AT.% SI

1: 41 AT.% SI
2: 66 AT.% SI

$\overline{C}$ = 56 AT.% SI

Figure 5:    (a) Thermal history of a sample quenched immediately after the primary phase nucleation at 55 K supercooling.   (b) General appearance of the microstructure.   (c) Detail of γ plate.   (d) Eutectic matrix.

the general appearance of the cellular eutectic microstructure.   Detail of the fine coupled eutectic at the cell centers is shown in Figure 6(d), while coarse eutectic at the cell boundaries is shown at a higher magnification in Figure 6(e).   The cell centers have an average composition of 57 at.% Si, while the coarse cell boundaries have an average composition of 59 at.% Si.   These compositions indicate that the last liquid (cell boundaries) is slightly enriched in solute (Si).   The average composition at the cell centers indicates a shift in the eutectic composition.

The microstructures shown in Figure 6 are thought to result from rapid initial growth during recalescence of the liquid, leading to a very

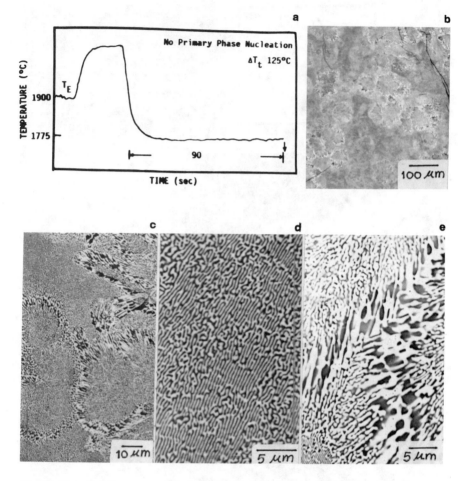

Figure 6: Sample supercooled to 120 K without primary phase nucleation.
(a) Thermal history showing a sample held in the supercooled state for 90
s. (b) General appearance of the microstructure. (c) Microstructure at
a higher magnification. (d) Detail of cell center. (e) Cell boundaries.

fine coupled eutectic microstructure. This is followed by growth at a
slower rate after recalescence due to smaller (or no) supercooling,
resulting in coarse coupled eutectic at the cell boundaries. An
interesting feature of the microstructural morphology in this sample is
the appearance of a quasi-regular coupled eutectic as opposed to the
irregular eutectic structure observed in the superheated samples or in
samples which exhibit primary phase nucleation. The appearance of a
quasi-regular eutectic microstructure, notwithstanding the facet forming
tendency of γ and δ, indicates coupled growth under extremum conditions.

The thermal history of a sample in which the first phase nucleation
occurred at 60 K below the liquidus, and quenched from a total
supercooling of 210 K, is shown in Figure 7(a), and the corresponding

186

microstructure at low magnification is shown in Figure 7(b). The microstructure is complex with γ plates surrounded by a thin broken layer of δ (not seen at this magnification). A coupled γ + ɛ̄ eutectic surrounds both the γ plates and the dendritic δ elements.

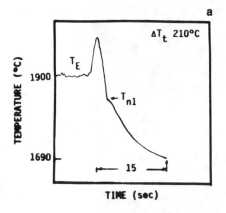

Figure 7: Sample quenched after a primary phase nucleation at 60 K, followed by additional supercooling of 150 K. (a) Thermal history. (b) General appearance of the microstructure.

Details of the microstructural morphology are shown in Figure 8(a) where γ plates are surrounded by a coupled eutectic zone, while the remaining matrix consists of dendritic δ with interdendritic eutectic, indicated by "1," "2," and "3," respectively. Note that the eutectic zone surrounding the γ plates and the interdendritic eutectic are continuous. A thin layer consisting predominantly of δ surrounds the γ plates as shown by the higher magnification micrograph in Figure 8(b). The eutectic zone near the γ plate is also seen in this figure, while the appearance of the matrix is shown in Figure 8(c).

Detail of the microstructure of γ plates in the above mentioned sample is shown in the three micrographs of Figure 9. Note the presence of a substructure in the plates together with regions which appear to be "untransformed." As discussed later, the substructure may be a result of a solid state eutectoidal decomposition of $\gamma_2$ to $\gamma_1$ and δ. The three dimensional microstructural morphology of these samples is shown in Figures 10(a) and (b). The micrographs, taken from a shrinkage cavity, show the dendritic δ matrix with coupled interdendritic eutectic of γ + δ. The strongly facet-forming γ appears either rod-shaped or plate-like. It can be seen that δ also exhibits a tendency towards faceting, although not as strong as γ.

Compositional Analysis

The average compositions of various phases and/or phase mixtures in samples with the microstructural morphology discussed above were measured using point-by-point line scanning in steps of 1 μm. The average composition was taken to be the arithmetic average of compositions from all points in the scan. Due to microstructural heterogeneity, large beam analysis was not used.

187

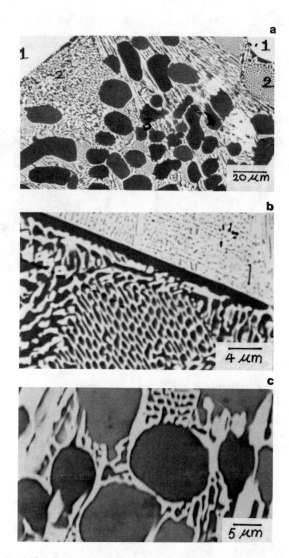

Figure 8:    Detail of microstructural morphology of sample shown in
Figure 7.   (a) General microstructure.   1 ≡ γ plates, 2 ≡ Coupled
eutectic, 3 ≡ Dendritic δ matrix with interdendritic coupled eutectic.
(b) γ plate surrounded by a thin δ layer.   (c) Detail of the dendritic δ
matrix and interdendritic coupled eutectic.

Typical results of compositional linescans across a eutectic near the
γ plates, of interdendritic eutectic in the matrix, and across dendritic
δ are shown together with their corresponding microstructures in Figures
11, 12, and 13, respectively.   Several points may be noted:   (1) the δ
composition is close to 65 at.% Si, (2) the composition of γ varies
between 40 and 45 at.% Si, (3) the average composition of the eutectic
near γ plates and in the interdendritic eutectic in the matrix is 54 to
55 at.% Si.   These compositional and morphological observations suggest

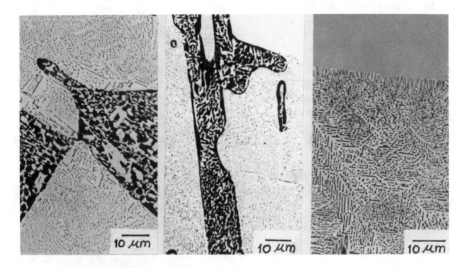

Figure 9: Microstructure of γ plates.

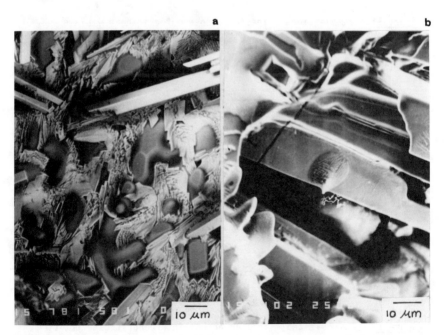

Figure 10: (a) Shrinkage cavity microstructure showing rod-shaped γ and dendritic δ with interdendritic coupled eutectic. Note the strong facet forming tendency of γ and δ. (b) Shrinkage cavity microstructure showing plate shaped γ.

that the eutectic zone around γ plates and the interdendritic eutectic are not two separate entities, but formed at the same time during the solidification sequence.

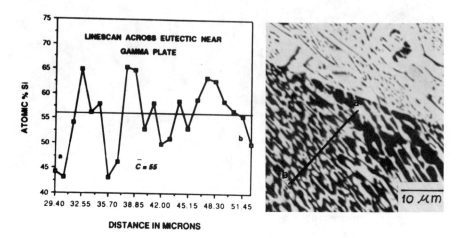

Figure 11: Compositional linescan across eutectic near γ plate.

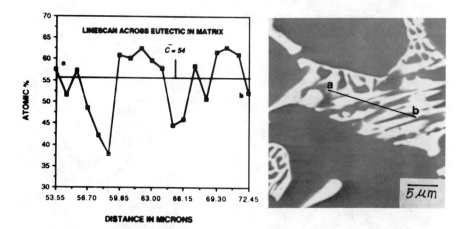

Figure 12: Compositional linescan across interdendritic eutectic.

Compositional analysis of a γ plate together with the corresponding microstructure is shown in Figure 14. The linescan shown in the figure was taken across the line a-c; the point b denotes the boundary between the single-phase and two-phase regions in the γ plate. From the linescan

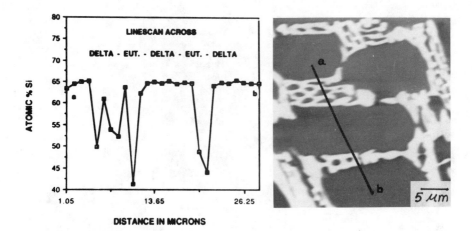

Figure 13: Compositional linescan across dendritic δ matrix.

and the corresponding micrograph, three important points may be noted:
(1) the average composition in the single-phase and two-phase regions is
nearly the same and close to 43 at.% Si, (2) the compositional variation
in the single-phase region is smaller than that in the two-phase region,
and (3) the darker (Si-rich) phase in the two phase region appears to be
rod-shaped: the cross sectional appearance is that of a collection of
rods with axial orientations predominantly in two mutually perpendicular
directions, yielding oval or cylindrical shaped features on the random
microsection. The last point is brought out clearly in the micrographs
presented in Figure 15, which show the detail of the single phase to two
phase transition interfaces. The highly oriented appearance of the
second phase also indicates a relationship with the crystallographic
symmetry of the plate.

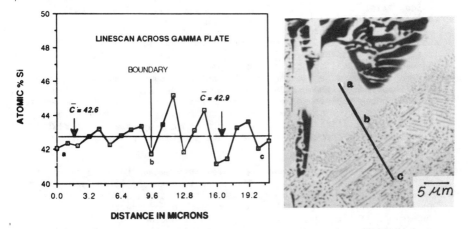

Figure 14: Compositional linescan across γ plate showing two phase and
single phase regions.

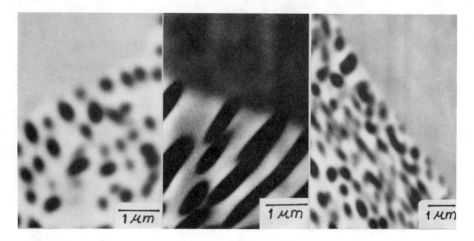

Figure 15: Detail of the interface between the single phase to two phase transition in γ plates.

The transformation from a single-phase to a two-phase region can occur by two entirely different processes:

(a) an abrupt change in the growth velocity of the faceted interface can lead to a breakdown in the solidification front, resulting in liquid entrapment of the type observed by (14) and (15). However, one consequence of liquid entrapment would be an increase in the average composition of the resulting two phase region (since the entrapped liquid would be richer in Si). Such an increase has not been observed in the present instance. Also, a sharp increase in the velocity of the growth front would be expected to lead to a morphological breakdown. Indeed, morphological instabilities have been observed in γ plates, as shown in Figure 16. These have an entirely different character and lead to branching, Figure 16(a), or liquid entrapment, Figure 16(b). These instabilities are likely to be due to the sensitivity of faceted growth to local solidification conditions.(16)

(b) A second mechanism by which the observed transition may be explained is eutectoidal decomposition. Evidence for the decomposition is strong. As discussed earlier, the work of (8) and present thermodynamic modeling both indicate the presence of a eutectoid reaction. Also, after a eutectoidal decomposition, the average composition would remain unchanged. Furthermore, since the $\gamma_2$ which forms from the supercooled liquid is supersaturated in Si with respect to its low temperature homogeneity limit, there is an added driving force for the decomposition of single-phase $\gamma_2$ into a two-phase $\gamma_1 + \delta$ mixture. In addition, the decomposition will leave the facets essentially unchanged.

Eutectoidal decompositions are typically slow due to low diffusivity in the solid state. However, in the present instance, it is likely that the release of the heat of fusion from solidification of the remaining liquid may sufficiently retard the cooling rate for the transformation to proceed. Evidently, due to kinetic reasons, some portions of $\gamma_2$ plates remain untransformed. As an extension of this reasoning, a higher rate of heat removal may retard or suppress the decomposition entirely. Our preliminary results on splat cooled samples support this finding.

192

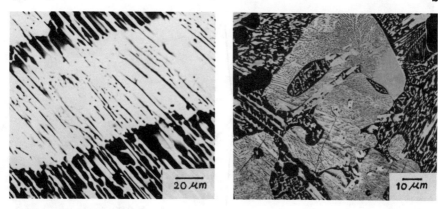

Figure 16: Morphological instability in $\gamma$: (a) Branching. (b) Liquid entrapment.

On the basis of these preliminary considerations, we believe that the substructure in $\gamma_2$ plates is the result of a eutectoidal decomposition into $\gamma_1 + \delta$, as shown in Figure 17. However, a more concrete description of this phenomenon must await a detailed TEM and x-ray analysis.

Solidification Sequence

From the compositional and morphological results described above, a solidification sequence may be deduced for the alloy upon supercooling of the liquid. An outline of the sequence is given first, followed by a detailed discussion of each factor involved. The solidification sequence is discussed with the aid of Figure 18 which shows the $\gamma$ and $\delta$ region of the equilibrium diagram. The metastable liquidus extensions of the $\gamma + L$ and $\delta + L$ regions and the metastable extensions of $\gamma$ and $\delta$ single phase fields are shown by dashed lines. The approximate boundaries of an asymmetric coupled eutectic growth zone (i.e., the region of temperatures and compositions in which the eutectic interface is stable) is also shown schematically by the central shaded region. The region is shown skewed towards the faceted intermediate phase, $\gamma$, following the reasoning of (4) on the shape of the coupled zone, indicating greater kinetic attachment difficulties during the growth of this phase.

Upon supercooling, $\gamma$ nucleates first in the liquid (point a), establishing a Si-enriched layer ahead of the interface and increasing the Si concentration in the liquid. The composition of $\gamma$ is given by the extended solidus line, as shown in Figure 18. As the liquid supercools further, the $\gamma$ nuclei continue to grow, until a certain supercooling and Si concentration are reached where $\delta$ of a composition given by the metastable extension of its phase field nucleates (point b). Rapid growth of $\delta$ phase follows its nucleation. This is due to the fact that once nucleated, the $\delta$-L interface is exposed to a large supercooling shown by $\Delta T(\delta)$, as compared to the supercooling at the $\gamma$-L interface. High amount of heat of fusion is released into the liquid, leading to an abrupt temperature rise (recalescence) which is also accompanied by a change in the composition of the liquid (point c). As shown in Figure 18, at this composition and temperature, the liquid is in the

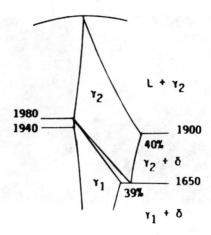

<u>Figure 17</u>:  Possible eutectoidal decomposition in γ.

coupled growth zone.  The liquid will then undergo stable coupled
eutectic growth, since both γ and δ are now present.  Also, following
recalescence, the solidification in the remaining liquid will proceed at
a slower rate, giving rise to a relatively coarse eutectic spacing.

The microstructural and compositional consequences of the above
description may be summarized as follows:  (1) γ should contain Si in
excess of the equilibrium value, (2) δ near the γ plates will be slightly
poorer in Si than the stoichiometric composition, and (3) a coupled
eutectic will surround both the solids and have an average composition
lower than $C_E$.

The solidification sequence described above underlines the important
distinction between solidification processing through bulk supercooling
vs. processing through directional solidification.  Generally, the
theories of coupled eutectic growth are based on steady state directional
solidification, where both phases are present during growth from the
liquid.  However, during bulk supercooling, the microstructural evolution
depends upon the hierarchy of nucleation of the constituents, as well as
the potency of heterogeneous nucleants present in the liquid.  Thus, for
example, primary γ phase nucleation can occur in the stability of regime
of an otherwise coupled eutectic zone leading to decoupled growth.

The growth kinetics of γ in the liquid are expected to be controlled
by the interface attachment kinetics (interface reaction controlled
growth) since γ is a highly faceted phase and since the heat and mass
transfer to and away from the interface are expected to have only a small
effect due to convection currents arising from electromagnetic stirring.

Figure 19 shows the change in the volume fraction of γ as a function
of the square root of time allowed for growth.  The volume fractions were
measured by quantitative microscopy of several samples.  The respective
growth times were measured from the recorded thermograms.  The growth of
γ follows a parabolic time dependence.  This appears to be rather
surprising since the dependence would indicate diffusion control.

194

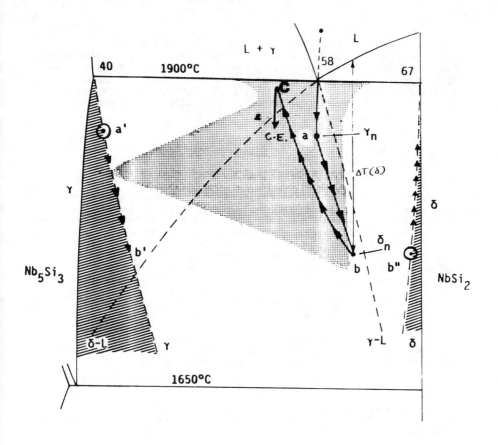

Figure 18: Solidification sequence in Nb-58 at.% Si alloys.

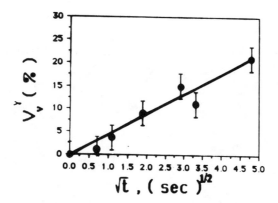

Figure 19: The volume fraction of $\gamma$ as a function of the square root of time allowed for growth during continuous cooling.

However, it must be noted that the growth of γ took place under continuous cooling conditions and not isothermally. This indicates that, under isothermal growth conditions, the growth of γ would be slower than that observed under the present conditions, indicating that γ growth kinetics are slower than those for diffusion control and may approach those for interface control. However, a more definitive answer must await further isothermal growth experiments.

## Summary

(a) During bulk supercooling, primary phase nucleation and growth play an important role in determining the growth morphology.

(b) The primary phase nucleation in the alloy occurred usually at a bulk supercooling of approximately 60 K. The primary phase was found to be the faceted intermediate phase γ. Upon further cooling, the remaining liquid continued to supercool by an appreciable extent, even in the presence of solid particles, until the second phase, δ, nucleated. The growth of δ was accompanied by a rapid recalescence. The final liquid to solidify formed a coupled eutectic with Nb content greater than the original eutectic composition.

(c) When the primary phase nucleation was suppressed, a fine cellular eutectic with quasi-regular morphology was obtained.

(d) Our preliminary results and phase diagram evaluation indicate the presence of a eutectoidal-type decomposition of γ. However, further study is required before reaching a definite conclusion.

## Acknowledgement

This research was supported by NASA-Vanderbilt Center for Space Processing of Engineering Materials. Special thanks are due to Professor Robert Bayuzick for his valuable discussions and input.

# References

1. V.S. Sudovtsova, G.I. Batalin, and V.S. Tutevich, Zh. Fiz. Khim., 59(9), 2156-2158 (1985).

2. G.V. Zemskov, R.L. Kogan, V.M. Lukyanov, and V.S. Viderman, Phys. Met. Metallogr., 25(4), 173-175 (1968).

3. M.S. Tsirlin, A.V. Kasatkin, and A.V. Byalobzheskii, Sov. Powder Metall. Met. Ceram., 17(12), 920-923 (1978).

4. W. Kurz and D.J. Fisher, Int. Met. Rev., (5-6), 177-204 (1979).

5. G.E. Amaya, J.A. Patchett, and G.J. Abbaschian, Grain Refinement in Casting and Welds (Proc. Conf.), 51-65, Fall Meeting of the Metallurgical Society of AIME, St. Louis, Missouri, October 25-26, 1982, Ed.: G.J. Abbaschian and S.A. David, Published: 1983.

6. W.G. Moffatt, The Handbook of Binary Phase Diagrams, Publ.: General Electric Company, Schenectady, New York (1979).

7. Metals Handbook, Volume 8, p. 283, 8th Edition, Publ.: ASM, Metals Park, Ohio (1973).

8. Yu. A. Kocherzhinskii, L.M. Yupko, and E.A. Shishkin, Russ. Metall., (1), 184-188 (1980).

9. G.V. Samsonov, Zh. Neorg. Khimii., 3(4), 868 (1958).

10. V.M. Pan, V.V. Petkov, and O.G. Kulik, Physics and Metallurgy of Superconductors, 183-187 (1970).

11. R.M. Waterstrat, K. Yvon, H.D. Flack, and E. Parthe, Acta. Cryst., (B31), 2765-2769 (1975).

12. I. Barin, O. Knacke, and O. Kubaschewski, Thermochemical Properties of Inorganic Substances, (Supplement), Springer-Verlag, New York, (1977).

13. A.B. Gokhale and G.J. Abbaschian, To be published.

14. G.J. Abbaschian and R. Mehrabian, J. Cryst. Growth, (43), 433-445 (1978).

15. S.D. Peteves, Growth Kinetics of Faceted Solid-Liquid Interfaces, Ph.D. Dissertation, University of Florida, (1986).

16. R. Elliott, Eutectic Solidification Processing, Crystalline and Glassy Alloys, Publ.: Butterworths & Co., Chapter 4 (1983).

# RAPID SOLIDIFICATION OF ALUMINA-ZIRCONIA

## EUTECTIC AND HYPOEUTECTIC ALLOYS

T. Whitney, V. Jayaram, C.G. Levi and R. Mehrabian

Materials Department
Department of Mechanical and Environmental Engineering
University of California
Santa Barbara, CA 93106

## Abstract

A technique has been developed to produce rapidly solidified fine pow-ders of binary ceramic alloys by electrohydrodynamic (EHD) atomization. Eu-tectic (42.5 wt%) and hypoeutectic (17 wt%) alumina-zirconia mixtures were prepared by colloidal chemistry methods, extruded and sintered to form thin rods which were subsequently atomized. Powders from tens of nanometers to the hundred micrometer range were produced in this manner. Scanning and transmis-sion electron microscopy of these materials revealed a variety of microstruc-tures including completely amorphous particles, single phase microcrystalline solid solutions and two phase mixtures of $Al_2O_3$ and $ZrO_2$. Although zirconia appeared predominantly in the monoclinic form, alumina exhibited a multipli-city of metastable crystal structures, including cubic spinel ($\gamma$), orthorhom-bic ordered spinel ($\delta$) and monoclinic ($\theta$). Microstructure evolution is dis-cussed in terms of composition and particle size, hence achievable supercool-ing, with particular emphasis on phase selection and solute redistribution.

Solidification Processing of Eutectic Alloys
D.M. Stefanescu, G.J. Abbaschian and R.J. Bayuzick
The Metallurgical Society, 1988

199

The development of rapid solidification processing (RSP) in ceramics has received limited attention even though ceramics are well known to exhibit more significant departures from equilibrium than metals during phase transformations. Adding to the technical complications of melting and handling liquid ceramics, there is a common misconception that RSP of these materials is not feasible due to their inherently low thermal conductivity [1]. Nevertheless, the higher structural complexity and slower atomic transport in most ceramics translate into considerably more sluggish solidification kinetics and thus permit rapid solidification even in the absence of rapid quenching. For example, there are many more glass-forming systems in ceramics than in metals, and the cooling rates necessary to avoid crystallization are generally much lower. It should then be possible to achieve significant supercoolings in liquid ceramics and open more avenues to microstructural control including refinement of the microstructural features, selection of metastable phases and extension of solid solubility.

Alumina-zirconia, an important alloy system in the development of tougher structural ceramics, is particularly attractive for fundamental solidification studies because its components exhibit numerous crystalline polymorphs, some of which could be formed and/or retained as metastable phases by RSP. Furthermore, the equilibrium phase diagram shown in Figure 1 reveals a rather simple liquidus curve with a single eutectic reaction and limited solid solubility of the components in each other.

Aluminum oxide exhibits only one stable crystalline form, corundum $(\alpha)$, consisting of a hexagonal close packed array of oxygen anions with the aluminum cations occupying two thirds of the octahedral interstices in order to preserve charge neutrality. It is well known, however, that numerous metastable forms of $Al_2O_3$ can be produced by calcination of hydroxides [3], vapor

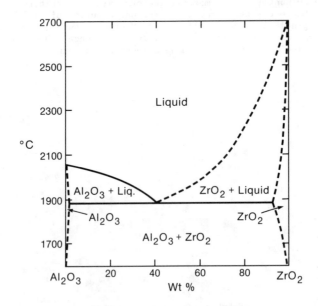

Figure 1 - $Al_2O_3$-$ZrO_2$ equilibrium phase diagram [2].

deposition [4], and rapid solidification techniques [1]. These include cubic
($\gamma$), orthorhombic ($\delta$) and monoclinic ($\theta$) structures based on the FCC oxygen
packing characteristic of spinel. The differences between them arise from the
configuration of the cation sublattice.

Contrary to alumina, zirconia exhibits three polymorphs which are stable
in the following temperature ranges [5]

| | 1443 K | | 2643 K | | 2953 K | |
|---|---|---|---|---|---|---|
| monoclinic (m) | $\rightarrow$ | tetragonal (t) | $\rightarrow$ | cubic (c) | $\rightarrow$ | liquid |

c-$ZrO_2$ has the crystal structure of fluorite, with the cations in an fcc arrangement and the anions occupying the tetrahedral interstices[1]. The structure is a crystallographic oddity because the cation/anion radius ratio is
too small to produce the coordination 8:4 characteristic of fluorite but rather large to fit comfortably in the 6:3 coordination typical of rutile. The
t and m forms of $ZrO_2$ are basically distorted versions of the fluorite structure, suggesting that covalency in the Zr-O bond plays a significant role in
maintaining a higher coordination number than that dictated by the relative
ion sizes [6]. The structural similarity makes the transformations between
these polymorphs rather easy and hinders the metastable retention of the high
temperature forms upon quenching in the absence of stabilizing additions,
e.g. $Y_2O_3$, CaO or MgO.

Rapid solidification studies in $Al_2O_3$-$ZrO_2$ have focused primarily on the
eutectic composition, $\sim$42 wt% $ZrO_2$. Amorphous structures are readily produced
by plasma spraying [7,8], shock-wave quenching (SWQ) [9] and melt extraction
[10], all of which involve solidification on a metal substrate, as well as in
powders smaller than 80 $\mu$m produced by flame pressure atomization (FPA) [9].
Analysis of coarser FPA powders reveals microcrystalline structures consist-
ing of $\epsilon$-$Al_2O_3$[2] + t-$ZrO_2$ below 160 $\mu$m and lamellar growth of $\alpha$-$Al_2O_3$ + t-$ZrO_2$
above this size, with partial transformation of tetragonal to monoclinic $ZrO_2$
above 200 $\mu$m. Coupled eutectic growth between $\delta$-$Al_2O_3$ and t-$ZrO_2$ has also
been reported in melt extracted flakes [11]. Spacings between 15 and 100 nm
were measured in hammer-and-anvil eutectic splats [8], and associated with
growth velocities up to 5 cm/s based on relationships established by directional solidification [12].

Melt extraction of off-eutectic compositions in the $Al_2O_3$-$ZrO_2$ system
has been reported to yield dendrites of the equilibrium primary phase
($\alpha$-$Al_2O_3$ or t-$ZrO_2$) surrounded by an interdendritic constituent which may be
a second phase segregate (t-$ZrO_2$), lamellar eutectic or a glassy phase [11].
Compositions up to 15 wt% $ZrO_2$ in $\delta$-$Al_2O_3$ and 12 wt% $Al_2O_3$ in t-$ZrO_2$ were
measured by EDS in these samples and ascribed to solute trapping during solidification, although the solid solution precipitates the excess solute in
fine crystalline form upon cooling [11]. Another important observation is
that $ZrO_2$ is retained in the t-form even though $Al_2O_3$ is not known as a stabilizer of any of the $ZrO_2$ polymorphs.

It is well acknowledged that atomization is the RSP process with the
highest potential for achieving significant supercoolings in the liquid,
hence departures from equilibrium and metastable microstructures [13]. Recent

---

1    Since the anions are larger, the structure may be defined more properly
as a simple cubic array of oxygen anions with half of the interstices
occupied by Zr cations.

2    $\epsilon$-$Al_2O_3$ is a metastable hexagonal phase reported to form by RSP. Its
identity and structure are still a subject of debate.

developments in electrohydrodynamic atomization extended the process to ceramic materials like $Al_2O_3$ which are insulators at room temperature but have a significant electrical conductivity near their melting points [14]. The EHD process was used in the present investigation to produce $Al_2O_3$-$ZrO_2$ powders ranging in size from a few nanometers to hundreds of microns. While not of commercial importance, these powders are particularly suitable for fundamental solidification studies as has been demonstrated with metallic materials [15]. The wide range of particle sizes results in a large spectrum of supercoolings and provides an excellent opportunity to study the effects of the latter on microstructure evolution and phase selection.

## Experimental

The developments leading to the application of electrohydrodynamic atomization to ceramics have been described in detail elsewhere [14]. Briefly, the containerless source developed for metals [16] is modified in the manner depicted in Figure 2. A ceramic rod, 1-2 mm in diameter, is passed through a small tantalum spring which is in electrical contact with both the sample and a +30 kV DC power supply. An electron beam is generated by thermionic emission from a heated filament at zero mean voltage. The beam is attracted to the highly positive spring/rod assembly and focused with a 500-700 V negative voltage applied to the filament grid. At the start, while the ceramic is electrically insulating, the beam heats up the tantalum spring which in turn heats up the sample rod. As the temperature of the rod approaches its melting point, the electrical conductivity increases to the point where it can sustain the beam current, resulting in melting of the rod tip. Under the intense electric field, the small liquid meniscus adopts the shape of a Taylor cone [17] whose apex becomes hydrodynamically unstable and breaks into a beam of microdroplets. The powders may be allowed to solidify in flight or splatted onto a copper chill placed in front of the source. The fine powders are typically collected on copper grids with carbon films, whereas the coarser ones are gathered as loose particulate from the bottom of the chamber.

## Preparation of Rod Feedstock

Once the problem of electrical conductivity was solved, the main hindrance to the production of ceramic alloy powders by EHD atomization was the production of rod feedstock of the desired composition. This was accomplished using a colloidal technique based on the approach suggested by Carlstrom and Lange [18].

High purity powders of $Al_2O_3$[1] and $ZrO_2$[2] were mixed separately in distilled water to yield slurries with 4 - 5 volume percent solids. The particles were then dispersed into stable suspensions by adding HCl until a pH of 2 was reached and further mixed for approximately two minutes with a small homogenizer. The suspensions were sedimented overnight during which time particles and agglomerates larger than ~1 μm settled to the bottom. The dispersions containing particles < 1 μm were then decanted and flocculated by adding $NH_4OH$ to adjust the pH to 8.

The solid content of the two slurries was determined by density measurements and appropriate amounts of each were then weighed out to form the desired composition. The slurries were mixed and homogenized, adding 2 - 4 wt% of polyvinyl alcohol as a binder. This mixture was then put into a 20 cm$^3$ syringe and centrifuged for 20 min at 2800 rpm and a relative centrifugal force of 1350 g to densify the slurry into a thick paste. The latter was extruded

---

1     Reynolds High Purity Alumina (99.98+), RC-HP DBM, Lot BM-1098.
2     Alfa Products, 99+% pure, 1-3 μm.

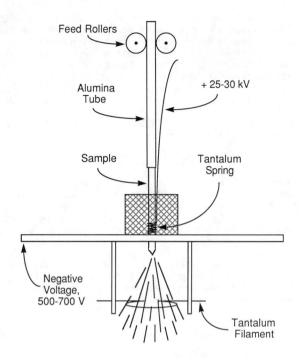

Feed Rollers

Alumina
Tube

+ 25-30 kV

Sample

Tantalum
Spring

Negative
Voltage,
500-700 V

Tantalum
Filament

Figure 2 - Schematic of the electrohydrodynamic atomization
source modified for the production of ceramic powders [14].

through the syringe into rods, 2 mm in diameter and 5 - 10 cm long, which
were dried in air overnight. They were then placed in a furnace and heated
slowly to 870 K to burn out the binder and subsequently sintered at 1870 K
for one hour. The alloy compositions were hypoeutectic $Al_2O_3$-17 wt% $ZrO_2$ and
eutectic $Al_2O_3$-42 wt% $ZrO_2$.

Analytical Techniques

Since the amount of particulate produced in each alloy was quite small,
the analysis was limited to scanning and transmission electron microscopy;
X-ray diffraction is underway as more material becomes available. In spite of
the small quantities, coarse loose powders were size-separated in three fractions: >50 μm, 25-50 μm and <25 μm.

Sections for SEM analysis were prepared by mounting powders in resin and
polishing with diamond paste. No etching was done in order to avoid any preferential leaching that could affect the microchemical analysis. SEM analysis
was carried out in a JEOL-840 equipped with a Tracor-Northern EDS system.

Powders smaller than 400 nm in diameter are electron transparent at
200 kV and were examined in the TEM as collected on the carbon films. TEM
specimens of the coarser powders were prepared by mixing them with epoxy,
centrifuging in a pointed container to concentrate them in a small volume,
curing the epoxy, dimpling and ion milling [19]. Owing to the small quantities available, the sample yield was erratic but allowed about 20-30 parti-

cles below 50 μm to be examined in each alloy. Thin areas were invariably
seen only at the periphery of the coarse particles. Since the dimpled thickness was ∿40 μm, it should be appreciated that prior to milling, all particles below 80 μm in diameter would be ground no further than their original
diameter. Given the fact that only the edge of the ceramic powder is thinned
before the surrounding epoxy is completely sputtered away, the projected
shadow of the powder is assumed to be its true diameter. Thus, the particle
size could be related directly to microstructure. Analytical TEM and micro-
diffraction were carried out in a JEOL 2000 FX. Images from thick regions
were obtained using a Kratos 1.5 MeV HVEM at the National Center for Electron
Microscopy, Lawrence Berkeley Laboratory.

## Phase Selection in Pure Alumina

The hierarchy of phases produced in the EHD atomization of pure $Al_2O_3$
serves as a useful foundation to understand the structures that emerge in the
binary system. Earlier papers [14,20] revealed that powder microstructures
may reflect both phase selection during solidification and solid state transformations during recalescence and subsequent cooling. The phases observed
are related to particle size as shown in Table I.

As expected, large powders (>15 μm) exhibit the equilibrium structure of
corundum (α), whereas very fine powders (<100 nm) are completely amorphous. A
number of metastable crystalline phases are observed in the intermediate size
range, all based on the FCC oxygen ion arrangement characteristic of spinel.

Table I. Phase Selection Hierarchy as a Function of Particle Size[1]

| Size (μm) | % $ZrO_2$ | 0.01 | 0.10 | 1.0 | 10 | 100 |
|---|---|---|---|---|---|---|
| Amorphous | 0 | ←----------- | | | | |
| | 17 | ←--------------- | | | | |
| | 42 | ←--------------------- | | | | |
| Cubic spinel (γ) | 0 | | | ---------------------- | | |
| | 17 | | | ------=-=-=========== | | |
| | 42[2] | | | | | |
| Orthorhombic (δ) | 0 | | | | ---------- | |
| | 17 | | | | ========= | |
| | 42[2] | | | | | |
| Monoclinic (θ) | 0 | | | | ---------- | |
| | 17 | | | | ========= | |
| | 42 | | | ------=-=-===================== | | |
| Corundum (α) | 0 | | | | | ----------→ |
| | 17 | | | | | ========→ |
| | 42 | | | | | ======→ |

[1] (-----) Single phase, (=====) Two phase.
[2] Cubic (γ) and orthorhombic (δ) spinels were not observed in the eutectic
powders.

The most common phase is γ, which has the cubic structure of conventional spinel with 32 oxygens and 21 1/3 cations statistically distributed among the interstitial sites [21][1]. Another structure frequently found is δ, which is an ordered form of γ with an orthorhombic unit cell containing 96 anions and 64 cations [20]. The crystallographic relationship between γ and δ structures is illustrated in Figure 3, where $a_\delta = a_\gamma$, $b_\delta = 1.5 \ a_\gamma$ and $c_\delta = 2 \ a_\gamma$; note that the unit cell is tripled during ordering to produce an integer number of cations. Less frequent in pure $Al_2O_3$ is θ, which has a monoclinic structure isomorphous with β-$Ga_2O_3$ [24] and a unit cell containing 12 anions and 8 cations equally distributed among octahedral and tetrahedral sites. Its relationship to γ is also shown in Figure 3.

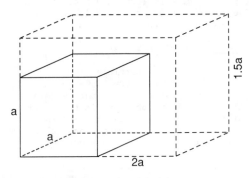
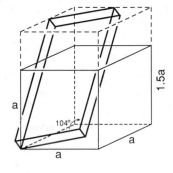

Spinel (γ)/orthorhombic (δ)          Spinel (γ)/Monoclinic (θ)

Figure 3 - Crystallographic relationship between the unit cells of the three metastable phases of $Al_2O_3$ based on an FCC array of oxygen anions. The lattice parameter of pure γ-$Al_2O_3$ is $a_\gamma$ = 0.79 nm.

Cubic spinel (γ) forms only from the liquid[2], but may easily undergo a phase transformation to more stable structures if exposed to elevated temperatures [14,25]. Calculations for pure $Al_2O_3$ reveal that sub-micron EHD powders are expected to retain γ even if nucleation occurs at relatively low supercoolings because the heat extraction rate is sufficiently fast to minimize the time exposure to temperatures where the solid state transformation kinetics is active [14]. On the other hand, powders above 10 μm would not be expected to retain γ, even if selected from the liquid, because they experience a much longer high temperature excursion. The experimental observations agree reasonably well with these predictions and show that powders between 1 and 10 μm exhibit a gradually increasing degree of transformation to δ. The transformation involves only cation rearrangements as indicated by particles that display symmetry related variants of δ and antiphase domain boundaries characteristic of a high → low symmetry ordering transformation.

---

[1]  The cubic spinel is labeled γ to conform to the overwhelming majority of the melt-spraying literature [22]. When the spinel is produced by calcination of hydroxides, the cubic form is conventionally known as η and the label γ is given to a tetragonally distorted variant. It is not clear, however, whether this tetragonal form is a separate phase or it is due to the presence of residual moisture in η [23].

[2]  Cubic spinel is also produced by devitrification of amorphous $Al_2O_3$ EHD powders under beam heating in the TEM.

Even though a major fraction of the powders showing the orthorhombic spinel structure (δ) are believed to originate from solid state ordering of γ, this phase can also form directly from the liquid as indicated by particles consisting of δ single crystals or multiple grains with high angle boundaries. On the other hand, the experimental observations indicate conclusively that θ does not evolve as an ordering product of γ or δ, but the presence of numerous defects suggests that the original solidification structure may also have undergone a solid state transformation [14].

All three phases are metastable relative to α-$Al_2O_3$ but the transformation involves a reconstruction of the anion sublattice--a more difficult process than that of ordering interstitial cations. It is estimated from calculations of the thermal history that some powders should exhibit microstructures representing intermediate stages in the transformation from the FCC to the HCP oxygen arrangement. No evidence of such a transformation was obtained, however, in spite of extensive microstructural analysis.

### Single Phase Alumina-Zirconia Powders

Phase selection in $Al_2O_3$-$ZrO_2$ as a function of particle size is conveniently summarized and compared with the results for pure $Al_2O_3$ in Table I. While zirconia was largely monoclinic where present, the primary alumina showed the same spectrum of crystallographic forms observed in the single component powders. However, the limits of the size ranges in which the different phases are observed are in general displaced to larger particles. The observations may be divided into single and two phase microstructures for the purposes of this discussion.

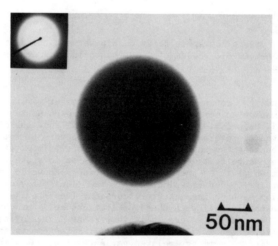

Figure 4 - TEM view of an amorphous particle in the $Al_2O_3$-17wt%$ZrO_2$ hypoeutectic alloy. The powder diameter is 225 nm.

Adding $ZrO_2$ to $Al_2O_3$ extends the size range in which amorphous powders, such as that in Figure 4, are observed. For example, the largest particle size in which amorphization was achieved is 100 nm for pure $Al_2O_3$ but increases to 300 nm in $Al_2O_3$-17 wt% $ZrO_2$ and about 1 μm at the eutectic composition.

206

Amorphization of a liquid droplet under continuous cooling requires the following condition to be satisfied

$$\int_{t_M^s}^{t_g} J^s(T,r_o) \, dt < 1 \tag{1}$$

where $J^s(T,r_o)$ is the rate at which nuclei of phase s form in a droplet of radius $r_o$ at temperature T, $t_M^s$ is the time at which the droplet crosses the melting temperature of phase s and $t_g$ is the time wherein the droplet reaches the glass transition temperature. Equation (1) must be satisfied for all possible crystalline phases (s = $\alpha$, $\gamma$, $\delta$, $\theta$, c, t, m, etc).

Following the conventional treatment of classical nucleation theory [26], the rate of nuclei formation in a droplet may be expressed as:

$$J(T,r_o) = n_a \kappa \exp \{-[(16\pi/3) \, (\Omega/\Delta G_{LS})^2 \, (\sigma\phi)^3]/(k_B T)\} \tag{2}$$

where $n_a$ is the number of molecules available for nucleation, equal to the total number of molecules in the droplet in the case of homogeneous nucleation and to the number of molecules in contact with a catalytic surface in the case of heterogeneous nucleation; $\kappa$ is a factor representing the attachment frequency of liquid molecules to the critical embryo, typically on the order of 1/picosecond for monatomic species and simple molecules; $\Omega$ is the molar volume; $\sigma$ is the solid-liquid interfacial energy, and $\phi$ is a geometric factor which represents the activity of the nucleation catalyst [14,27].

Nucleation in a binary system is driven by the change in free energy, $\Delta G_{LS}$, when an infinitesimal amount of solid of composition $X_S$ is formed from a liquid of composition $X_L$ [28], given as

$$\Delta G_{LS} = (1 - X_S) \, (\mu_S^A - \mu_L^A) + X_S \, (\mu_S^Z - \mu_L^Z) \tag{3}$$

where $\mu_j^i$ is the chemical potential of component i (A = $Al_2O_3$, Z = $ZrO_2$)[1] in the phase j (liquid and the relevant crystalline phase).

Increased glass forming tendency must arise as a consequence of slower nucleation kinetics of all the relevant crystalline phases and/or a higher glass transition temperature, hence a shorter time to cool from $T_M$ to $T_g$. One could argue on the basis of Equation (2) that a reduced nucleation frequency may result from an increase in the effective interfacial energy ($\sigma\phi$) or a higher viscosity of the supercooled liquid arising from a higher $T_g$. It is likely, however, that the largest effect of solute addition is on the driving force for nucleation, Equation (3).

The first term in Equation (3) represents the contribution of the solvent to the driving force and is normally negative at temperatures below the liquidus. When nucleation occurs below the $T_0$ curve[2] and above the solidus, the solute increases its chemical potential during solidification. Since the second term in Equation (3) would then be positive, the overall driving force for nucleation would be reduced. For systems where the metastable solidus

---

[1]  A more rigorous expression for the free energy should contain three terms since there are three ionic species in the liquid. However, the formulation given is adequate for the present discussion.

[2]  The $T_0$ curve is the locus of temperature-composition points wherein the liquid and solid free energies are equal.

drops rapidly with increasing solute content, as in $Al_2O_3$-$ZrO_2$, solidification below $T_0$ would always imply solute trapping and thus a reduced driving force. As the composition of the solid increases, the negative contribution of the solvent term is diminished and the positive contribution of the solute term is enhanced, further reducing $\Delta G_{LS}$ and J.

Amorphization in eutectic systems has been extensively discussed by Boettinger [29]. In general, coupled multiphase growth involves solute redistribution which must be effected by diffusion. If the solidification rate is increased to a point in which diffusion cannot occur in the time scale of the process, the two phase growth should be superseded by partitionless growth of a single phase. However, if the $T_0$ temperature is below the glass transition temperature, the liquid would be transformed to an amorphous solid before it reaches the point in which partitionless solidification is allowed by thermodynamics. It has been suggested [11] that the $Al_2O_3$-$ZrO_2$ system falls into this category since the shape of the phase diagram suggests that the $T_0$ curves for the terminal phases drop rapidly with composition.

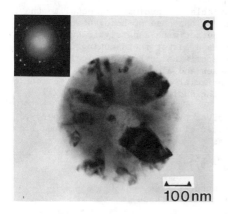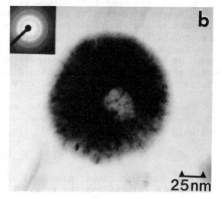

Figure 5 - Single phase microcrystalline powders of (a) cubic spinel $\gamma$ in the $Al_2O_3$-17 wt% $ZrO_2$ hypoeutectic alloy and (b) monoclinic $\theta$ in the $Al_2O_3$-42 wt% $ZrO_2$ eutectic alloy. No evidence of $ZrO_2$ segregates or precipitates was found by electron diffraction. Grain sizes are on the order of 20-40 nm for $\gamma$ and <5 nm for $\theta$.

Analysis of many sub-micron powders in both alloy compositions revealed that a small number of single phase microstructures were detected in the upper end of the size range where amorphous powders are observed. The crystalline phases are cubic spinel ($\gamma$) in the case of the hypoeutectic alloy and monoclinic $\theta$ in the eutectic composition, see Figure 5. Microchemical analysis revealed that these powders have the original bulk composition and showed no evidence of phase separation. This is an indication that solidification took place below the corresponding $T_0$ curves for both $\gamma$ and $\theta$, but above their $T_g$ temperature. Nevertheless, the glass transition temperature should be reasonably close to $T_0$ for both compositions, as indicated by the fact that single phase crystalline and amorphous powders exist in the same size range. In a random scale of events, some powders which are supercooled below $T_0$ would nucleate above $T_g$ whereas others may be supercooled down to the amorphous range.

Supersaturated single-phase $Al_2O_3$-$ZrO_2$ powders such as those in Figure 5 are commonly microcrystalline. This follows from the high supercoolings achievable in sub-micron powders, where the driving force for nucleation is high but the diffusion-controlled growth rates are increasingly sluggish as the droplet approaches $T_g$. It has been shown that these conditions lead to massive nucleation and microcrystallinity before any significant growth and the ensuing recalescence can take place [27].

It is not apparent why $\gamma$ gives way to $\theta$ with increasing $ZrO_2$ content both in the single and two phase microstructures, as listed in Table I. One could speculate that the incorporation of $ZrO_2$ in the spinel requires the creation of more cation vacancies in a structure that already possesses unoccupied cation sites. It is thus possible that the system adopts a more stable ordered form where the stoichiometry of the unit cell does not involve random vacancies after a certain solute concentration is reached.

## Two Phase Hypoeutectic Powders

Two phase structures of metastable $Al_2O_3$ in combination with m-$ZrO_2$ start to appear in hypoeutectic powders above $\sim$1 μm in diameter, see Table I. While $\gamma$ is the dominant phase in the low end of this size range, it gradually yields to $\delta$ and $\theta$ as the particle size increases. As in pure $Al_2O_3$, $\delta$ may evolve from the ordering of $\gamma$ or from the liquid, but $\theta$ or a related precursor appears to form independently of $\gamma$. In addition, it was qualitatively ascertained that the proportion of $\theta$ relative to $\delta$ was higher than that in pure $Al_2O_3$. Furthermore, the onset of the stable primary $\alpha$-$Al_2O_3$ was also displaced from $\sim$15 μm to $\sim$25 μm.

Typical solidification morphologies of coarse (>10 μm) $Al_2O_3$-17 wt%$ZrO_2$ powders are depicted in Figure 6. Powders above $\sim$25 μm in diameter exhibit a dendritic structure with average secondary arm spacings on the order of 1-2 μm, Figure 6(a). This is quite different from the structure characteristic of coarse pure $Al_2O_3$ powders, which are primarily faceted [14]. Furthermore, where dendritic growth is observed in pure $Al_2O_3$, the dendrite arms appear orthogonal and well developed, in contrast with the finer and more irregular appearance of the dendrites in Figure 6(a). The dendritic morphology gives way to a "cellular" structure in powders below $\sim$25 μm, see Figure 6(b), although the scale of the primary phase is quite similar in both cases.

Metallographic sections of coarse hypoeutectic powders reveal two distinct morphologies also shown in Figure 6. The microstructure of Figure 6(c) is thought to correspond to the dendritic morphology, whereas that in Figure 6(d) is associated with the cellular powders. Another significant difference is the morphology of the interdendritic constituent, corresponding in principle to liquid of eutectic composition. This appears as a divorced second phase in the majority of the dendritic powders, Figure 6(e), but often exhibits a well developed lamellar morphology, Figure 6(f), in the cellular powders and some of the coarser dendritic microstructures. EDS analysis in the SEM shows the divorced segregate to contain $\sim$94%$ZrO_2$ and $\sim$6%$Al_2O_3$, close to the equilibrium composition at the eutectic temperature. On the other hand, the average composition of the lamellar structure is quite close to the eutectic composition, $\sim$42 wt% $ZrO_2$.

TEM analysis of hypoeutectic powders was primarily focused on powders below 50 μm in diameter as it was expected that metastable effects were minimal above this size. Only microstructures containing the divorced segregate have been studied in detail because the sampling statistics have not yet produced a thin region in a powder with the interdendritic lamellar constituent. The divorced $ZrO_2$ segregate predominantly exhibits the twinned monoclinic structure shown in Figure 7(a), suggesting that it grew as t-$ZrO_2$ from the

209

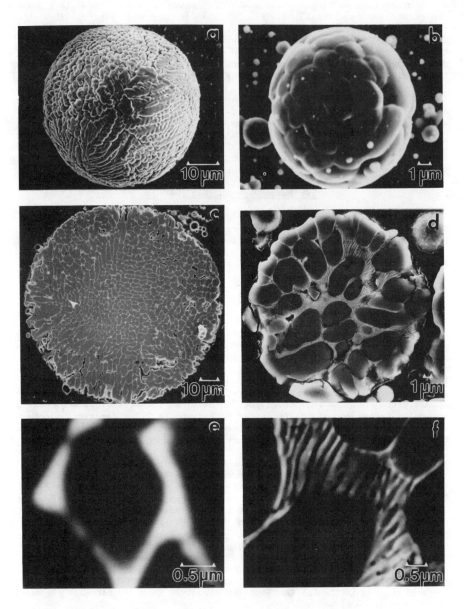

Figure 6 - Microstructures of coarse (>10 μm) hypoeutectic powders. (a) and (b) are SEM views of the dendritic and cellular surface morphologies, respectively; (c) and (d) are SEM views of metallographic sections corresponding to the structures in (a) and (b), but in different powders; (e) shows the divorced $ZrO_2$ segregate and (f) the coupled eutectic growth in the interdendritic spaces.

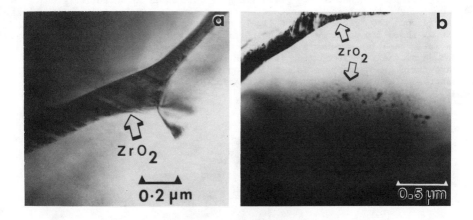

Figure 7 - TEM views of a dendritic microstructure in a ∿50 μm
powder showing (a) the interdendritic $ZrO_2$ second phase with a
twinned monoclinic structure and (b) fine $ZrO_2$ precipitation in
the α dendrite core and a PFZ next to the segregate.

liquid and transformed to m-$ZrO_2$ during cooling. This is also consistent with
the $Al_2O_3$ concentration determined by EDS, which is the maximum equilibrium
solubility of $Al_2O_3$ in the t-$ZrO_2$ phase predicted by the phase diagram.

The m-$ZrO_2$ segregate was found in combination with primary corundum in
particles larger than ∿25 μm and with δ or θ-$Al_2O_3$ below this size. The primary phase often exhibits very fine Zr-rich precipitates and a precipitate-
free zone (PFZ) near the dendrite/segregate interface, see Figure 7(b). The
average concentration of $ZrO_2$ in the α-$Al_2O_3$ dendrite core is in the range of
1 to 3wt%, slightly above the maximum equilibrium solubility. Furthermore,
preliminary analysis of powders where the primary phase is δ or θ instead of
α does not reveal a significant difference in $ZrO_2$ content in these phases.
It is not clear at this point whether the precipitation at the dendrite cores
is indicative of solute trapping during rapid solidification or it is merely
evident because the dissolved $ZrO_2$ in the PFZ precipitates preferentially on
the existing interdendritic phase as its solubility decreases during cooling.

Finally, many particles less than ∿10 μm in diameter appeared feature-
less on the surface, whereas those in pure $Al_2O_3$ revealed grain boundaries
indicative of an equiaxed structure [14]. In spite of the external feature-
less appearance, TEM analysis reveals that all hypoeutectic powders between 1
and 10 μm are two-phase, although the primary phase is now δ or θ and the
amount of interdendritic $ZrO_2$ is much less than that in the larger powders.

<u>Two Phase Eutectic Powders</u>

Phase selection in the eutectic composition follows the general trends
observed in pure $Al_2O_3$ and $Al_2O_3$-17wt%$ZrO_2$, as shown in Table I. However, a
single metastable phase, θ, predominates in the intermediate size range, suggesting that γ and δ are destabilized by the addition of $ZrO_2$ as discussed
above.

Two phase structures based on θ-$Al_2O_3$ start to appear in powders above
∿500 nm. In the lower end of the size range, the powders exhibit an irregular

211

array of $\theta$-$Al_2O_3$ and m-$ZrO_2$ grains, e.g. Figure 8, as it is unlikely that coupled growth will have enough time to develop. It is also believed that tetragonal $ZrO_2$ grows from the liquid and later transforms to the monoclinic form. The $\theta$ + m domain extends to particles as large as ~50 µm at which size a fully lamellar structure is present. However, owing to a paucity of finer powders, the transition from irregular to lamellar structures could not be ascertained. The $\theta$ phase is rapidly superseded by $\alpha$ as the powder diameter increases above 50 µm but the $ZrO_2$ remains in monoclinic form.

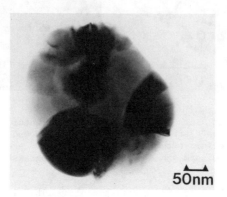

Figure 8 - TEM view of a non-lamellar two-phase eutectic powder containing $\theta$-$Al_2O_3$ and m-$ZrO_2$. This microstructure follows the single phase microcrystalline in Figure 5(b) as particle size increases.

Figure 9 shows an example of a well developed lamellar eutectic powder in the 100 µm size range. The eutectic cell size is on the order of 40 µm and the interlamellar spacing ranges from 100-300 nm. Since the phases growing from the liquid are probably $\alpha$-$Al_2O_3$ and t-$ZrO_2$ in the larger powders, one could use available directional solidification data to estimate the growth velocities. Stubican and Bradt [12] report the following relationship between spacing, $\lambda$, and growth velocity, V, for $Al_2O_3$-$ZrO_2$ eutectic:

$$\lambda^2 V = 10^{-17} \; m^3/s \tag{4}$$

One may thus deduce that solidification velocities in the range of 0.1 to 1 mm/s were achieved in these powders. These relatively low growth rates are not surprising since a 100 µm radiation-cooled droplet would only achieve a cooling rate on the order of $10^4$ K/s prior to solidification.

TEM work on eutectic powders revealed that both $\theta$ and $\alpha$-$Al_2O_3$ exhibit coupled lamellar growth with t/m-$ZrO_2$, as shown in Figure 10. Although most particles contain either $\alpha$ or $\theta$, occasionally both forms could be seen in the same powder, albeit in different colonies. The lamellar spacings did not vary significantly, but the $\alpha$ lamellae seem to be "pinched" at some points, as if they were breaking up to approach a rod morphology. Indeed, the $Al_2O_3$-$ZrO_2$ falls within the range of fibrous eutectics predicted by Hunt and Jackson [30], and both lamellar and fibrous morphologies have been observed experimentally [12,31], suggesting that the selection of a particular pattern may be affected by small changes in supercooling or growth velocity [12].

212

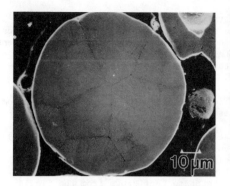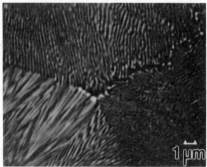

Figure 9 - Typical microstructures of coarse (>10 μm) $Al_2O_3$-$ZrO_2$ eutectic powders showing well developed lamellar growth in cells on the order of 40 μm. It is anticipated from the powder size that the phases present are α-$Al_2O_3$ and m-$ZrO_2$. Lamellar spacings are typically 100-300 nm.

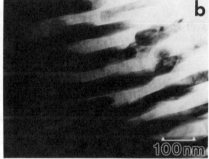

Figure 10 - Evidence of coupled growth between t-$ZrO_2$ (transformed to monoclinic) and α-$Al_2O_3$ (a) as well as θ-$Al_2O_3$ (b). Note that the morphology in (a) has a fibrous appearance while that in (b) is lamellar.

It may be useful as a final note to compare the microstructures produced by EHD, where achievable supercoolings are large but heat extraction rates are low, with those resulting from splat quenching [11], where cooling rates are high but supercoolings are limited in principle by the larger volumes of liquid and the heterogeneous nucleation on the chill surface. Lamellar spacings as fine as 15 nm have been reported in hammer-and-anvil splats, compared with 100-300 nm in this investigation. This would suggest achievable growth rates two orders of magnitude higher than those estimated above, i.e. up to ∿5 cm/s. Nevertheless, the splats invariably contained the stable α-$Al_2O_3$ while the larger supercoolings achievable in the powders produced the metastable form θ. On the other hand, t-$ZrO_2$ was retained in the splats, conceivably because the finer structure inhibited the t → m transformation due to the elastic constraint of the alumina matrix.

## Conclusions

Application of colloidal chemistry techniques and recent developments in EHD atomization offer a new avenue to study rapid solidification of ceramics of tailored compositions. The particle size distribution ranges from the tens of nanometers to the hundred micrometer range and provides a wide spectrum of achievable supercoolings, which can then be related qualitatively to microstructural observations.

Microstructures of $Al_2O_3$-$ZrO_2$ alloys processed in this manner vary from single phase amorphous and single phase microcrystalline to two-phase crystalline as the powder size increases. In addition to the stable corundum structure, the $Al_2O_3$ phase exhibits at least three distinct metastable crystalline forms, all based on the FCC anion packing characteristic of spinel. On the other hand, $ZrO_2$ is almost exclusively present as monoclinic, but there is significant evidence that this is a product of the solid state transformation from t-$ZrO_2$, which is the phase that grows from the liquid.

Increasing supercooling--by reducing particle size--changes the $Al_2O_3$ phase selection from more to less ordered phases, $\alpha \rightarrow \theta \rightarrow \delta \rightarrow \gamma$. However, some of the metastable phases selected at higher supercoolings may transform to more stable structures following recalescence during rapid solidification.

Increasing $ZrO_2$ content shifts the transition points in the phase selection hierarchy to larger particle sizes. Furthermore, $ZrO_2$ additions appear to favor the formation of $\theta$-$Al_2O_3$ over $\gamma$ and $\delta$-$Al_2O_3$.

The attainment of single phase ($\gamma$ or $\theta$) microcrystalline structures with a $ZrO_2$ content as high as 42.5wt% suggests that the $T_0$ curves for these two phases must lie above the glass transition temperature in the hypoeutectic range of the phase diagram.

## Acknowledgements

The authors are indebted to Dr. Frederick Lange for his advice on the development of the colloidal route to produce the ceramic alloys, to Dr. Juan Valencia for his technical assistance in the atomization experiments, and to the National Center for Electron Microscopy at the Lawrence Berkeley Laboratory for the use of the HVEM. This research was sponsored by the National Science Foundation under grant DMR-85-18985. The program director is Dr. Bruce MacDonald.

## References

1.  M.C. Brockway and R.R. Wills, "Rapid Solidification of Ceramics, A Technology Assessment", (Report MCIC-84-49, Metals and Ceramics Information Center, January 1984).

2.  A.M. Alper, "System $Al_2O_3$-$ZrO_2$ in Ar," in Science of Ceramics, vol. 3, ed. G.H. Stewart, (London, UK: Academic Press, 1967), 339.

3.  K. Wefers and G.M. Bell, "Oxides and Hydroxides of Aluminum", (Technical Paper No. 19, Alcoa Research Laboratories, 1972).

4.  A.L. Dragoo and J.J. Diamond, "Transitions in Vapor-Deposited Alumina from 300° to 1200°C," J. Am. Cer. Soc., 50 (11)(1967), 568-574.

5.  E.C. Subbarao, "Zirconia - An Overview," in Advances in Ceramics, vol. 3, Science and Technology of Zirconia, eds. A.H. Heuer and L.W. Hobbs, (Columbus, OH: American Ceramic Society, 1981), 1-24.

6.  A.H. Heuer and M. Rühle, "Phase Transformations in $ZrO_2$ Containing Cera-
    mics: I, The Instability of c-$ZrO_2$ and the Resulting Diffusion-Control-
    led Reactions," in Advances in Ceramics, vol. 12, Science and Technology
    of Zirconia II, eds. N. Claussen, M. Ruhle and A. Heuer, (Columbus, OH:
    American Ceramic Society, 1984) 1-13.

7.  A. Von Krauth and H.M. Meyer, Ber. Dtsch. Keram. Ges., 42 (1965), 61-67.

8.  G. Kalonji, J. McKittrick and L.W. Hobbs, "Applications of Rapid Solidi-
    fication Theory and Practice to $Al_2O_3$-$ZrO_2$ Ceramics," in Advances in
    Ceramics, vol. 12, Science and Technology of Zirconia II, eds. N. Claus-
    sen, M. Ruhle and A. Heuer, (Columbus, OH: American Ceramic Society,
    1984) 816-825.

9.  N. Claussen, G. Lindemann, and G. Petzow, "Rapid Solidification of
    $Al_2O_3$-$ZrO_2$ Ceramics," Mater. Sci. Monographs, 16 (1983), 489-497.

10. J. McKittrick, G. Kalonji and T. Ando, "Crystallization of Rapidly So-
    lidified $Al_2O_3$-$ZrO_2$ Eutectic Glass", submitted to J. Non-Cryst. Solids.

11. J. McKittrick, G. Kalonji and T. Ando, "Microstructural Control of $Al_2O_3$
    $ZrO_2$ Ceramics through Rapid Solidification", submitted for publication.

12. V.S. Stubican and R.C. Bradt, "Eutectic Solidification in Ceramic Sys-
    tems," Ann. Rev. Mater. Sci., 11 (1981), 267-297.

13. C.G. Levi and R. Mehrabian, "Fundamental Aspects of Rapid Solidification
    Processing," in Undercooled Alloy Phases, eds. E.W. Collings and C.C.
    Koch, (Warrendale, PA: The Metallurgical Society, 1986), 345-374.

14. C.G. Levi, V. Jayaram, J.J. Valencia and R. Mehrabian, "Phase Selection
    in Electrohydrodynamic Atomization of Alumina:, submitted to Journal of
    Materials Research.

15. C.G. Levi and R. Mehrabian, "Microstructures of Rapidly Solidified Alu-
    minum Alloy Submicron Powders," Metall. Trans. A, 13A (1982), 13-23.

16. J.F. Mahoney, S. Taylor and J. Perel, "Fine Powder Production using
    Electrohydrodynamic Atomization," (Paper presented at the IEEE/IAS An-
    nual Conference - Electrostatic Processes, October 1984).

17. G. Taylor, "Disintegration of water drops in an electric field," Proc.
    Roy. Soc. London, 280A (1964), 383-397.

18. E. Carlström and F.F. Lange, "Mixing of Flocced Suspensions," J. Am.
    Cer. Soc., 67 (8)(1984) C169-170.

19. M.J. Carr, "A method for preparing powder specimens for transmission
    electron microscopy," J. Elec. Micr. Tech., 2 (1985), 439-443.

20. V. Jayaram and C.G. Levi, "The Structure of $\delta$-Alumina Evolved from the
    Melt and the $\gamma \rightarrow \delta$ Transformation", submitted to Acta Metallurgica.

21. G. Ervin, "Structural Interpretation of the Diaspore-Corundum and Boeh-
    mite-$\gamma$-$Al_2O_3$ Transitions," Acta Cryst., 5 (1952), 103-108.

22. R. McPherson, "On the formation of thermally sprayed alumina coatings,"
    J. Mater. Sci., 15 (1980), 3141-3149.

23. A.M. Lejus, Rev. Int. Hautes Temper. et Refract., 1 (1964), 53.

24. S. Geller, "Crystal Structure of $\beta$-$Ga_2O_3$," J. Chem. Phys., **33** (3)(1960), 676-684.

25. R. McPherson, "Formation of metastable phases in flame- and plasma-prepared alumina," J. Mater. Sci., **8** (1973), 851-858.

26. J.W. Christian, The Theory of Transformations of Metals and Alloys, 2nd. Ed., Part I, (Oxford, UK: Pergamon Press, 1975), Chapter 10.

27. C.G. Levi, "Evolution of Microcrystalline Structures in Supercooled Metal Powders", Metall. Trans. A, **19A** (1988), in press.

28. J.C. Baker and J.W. Cahn, "Thermodynamics of Solidification," in Solidification, (Metals Park, OH: American Society for Metals, 1971), 23-58.

29. W.J. Boettinger, in Rapidly Solidified Amorphous and Crystalline Alloys, eds. B.H. Kear, B.C. Giessen and M. Cohen, (Amsterdam, NL: North Holland, 1982), p. 15.

30. J.D. Hunt and K.A. Jackson, "Binary Eutectic Solidification," Trans. AIME, **236** (1966), 843-852.

31. F. Schmidt and D. Viechnicki, J. Mater. Sci., **5** (1970), 470.

EFFECT OF COOLING RATE ON SOLIDIFICATION MICROSTRUCTURES

IN PSEUDOEUTECTIC 304SS-Zr ALLOYS

Z.S. Wronski and J.D. Boyd

Physical Metallurgy Research Laboratories, CANMET
Ottawa, Canada

Abstract

The effects of solidification rate on the phase transformations and
microstructures in rapidly solidified 304 stainless steel and 304 stainless
steel-11 a/o zirconium have been investigated.  Specimens prepared by arc
melting, melt spinning, and argon plasma sputtering were characterized by
optical- and electron metallography, X-ray diffraction, and Mössbauer
spectroscopy.  The solidification sequence and microstructure of 304 varies
with cooling rate.  304-Zr behaves as a pseudoeutectic for arc melting and
melt spinning.  Sputtered 304-Zr forms amorphous films.when the deposit
contains at least 9-12 a/o Zr, and sputtering is carried out under conditions
of low metal gas thermalization.

Solidification Processing of Eutectic Alloys
D.M. Stefanescu, G.J. Abbaschian and R.J. Bayuzick
The Metallurgical Society, 1988

## Introduction

It has been demonstrated that rapid solidification of Fe-base alloys can produce grain refinement, enhanced solubility of alloying elements and impurities, reduced segregation, and suppression of second-phase particles formed during secondary solidification or precipitation reactions (1,2). Rapidly solidified stainless steels show superior resistance to pitting corrosion, stress-corrosion cracking, and oxidation (3). Hence, there is interest in developing both microcrystalline and amorphous rapidly-solidified stainless steels for corrosion/oxidation barriers. Rapidly solidified stainless steels are usually microcrystalline. However, it was recently shown that amorphous films can be produced when 304 stainless steel is co-sputtered with zirconium to produce an alloy containing at least 10 atomic pct. Zr (4). The Fe-Zr binary system has a deep eutectic at 10 a/o Zr, and melt-spun ribbon of near-eutectic compositions of Fe-Zr alloys are amorphous (5,8). Zirconium also acts as a glass-forming element in Fe-Ni-Zr alloys (6).

The purpose of the present study was to investigate the phase transformations and microstructures in rapidly solidified 304 stainless steel ("304") and 304 stainless steel-11 a/o Zr ("304-Zr") as a function of solidification rate. In particular, clarification was sought on the solidification mechanisms of the 304-Zr alloy, which, in the absence of a phase diagram, we refer to as a "pseudoeutectic". To produce a range of solidification rates, the two alloys were produced by arc melting, melt spinning and argon plasma sputtering.

## Experimental

### Materials Preparation

The 304 starting material was from commercial stock. Arc-melted 100-g samples of this material and 304-Zr alloy were prepared on a water-cooled copper substrate. The system was first evacuated to $2 \times 10^{-5}$ torr and back-filled with argon to one atmosphere. To minimize segregation, the 304 samples were remelted twice, and the 304-Zr material was remelted four times. Samples for this study were taken from the area adjacent to the Cu substrate where the cooling rate is estimated to be $10^{+4}$ K/s. The compositions of the arc-melted alloys were determined by wet chemical analysis, atomic absorption spectroscopy, and Leco total oxygen analysis. The results are given in Table I.

Table I.  Experimental Material(arc melted, wt %)

|     | 304   | 304-Zr |
|-----|-------|--------|
| Cr  | 17.3  | 14.5   |
| Ni  | 8.8   | 7.7    |
| Mn  | 1.4   | 1.1    |
| Si  | 0.51  | 0.36   |
| Mo  | 0.26  | 0.20   |
| Cu  | 0.13  | 0.04   |
| Zr  | -     | 14.2   |
| C   | 0.042 | 0.044  |
| N   | 0.044 | 0.008  |
| O   | 0.015 | 0.0003 |

Melt-spun ribbon of both alloys was prepared from the arc-melted material by induction melting and single-roller casting in an argon atmosphere. The 304 alloy was cast at surface velocities of 15, 30 and 60 m/s, whereas the 304-Zr was produced at 30 m/s only. The resulting ribbon was 2-5 mm wide and 25-100 μm thick, depending on the casting speed. The typical cooling rate for this process is $10^{+6}$ K/s.

Two-10 μm-thick films of 304 and 304-Zr were prepared by RF argon plasma sputtering onto a polished copper substrate. The cooling rate for this process can be as high as $10^{+12}$ K/s (7). A limited number of films were deposited on aluminum foil or mica. The sputtering technique is described in detail elsewhere (4).

## Materials Characterization

Microstructures were characterized by X-ray diffraction (CoK$_\alpha$) and [57]Fe conversion electron Mossbauer spectroscopy (CEMS) from sample surfaces parallel to the substrate, and by optical- and scanning electron metallography of sections normal to the substrate. Metallographic samples of arc melted 304 were etched in modified Murakami reagent (30 g potassium hydroxide and 30 g potassium ferricyanide per 100 mL $H_2O$). Melt-spun 304 was electrolytically etched in chrome regia (1:1 by volume hydrochloric acid and chromic acid). Arc-melted and melt-spun 304-Zr were etched in Marble's reagent (4 g cupric sulphate and 20 mL hydrochloric acid per 20 mL $H_2O$). Sections through the sputtered films of both alloys were electropolished in 10% chromic acid, but it was not possible to avoid preferential etching of the Cu substrate. Transmission electron microscopy specimens were prepared of melt-spun 304 parallel to the ribbon surface by electropolishing in 5% perchloric acid in methanol followed by simultaneous ion milling from two sides.

Results

## X-Ray Diffraction

The X-Ray diffraction results are summarized in Table II. The arc-melted and melt-spun 304 exhibited strong fcc Fe-(111) and Fe-(200) peaks, indicating a predominately austenitic (γ) structure. The arc-melted samples and thickest melt-spun ribbon (50-100 μm) of 304 also exhibited weak bcc peaks, indicating the presence of some ferrite (δ) in these samples. The sputtered 304 films were approximately 2 μm thick and exhibited a strong bcc Fe-(110) peak as well as fcc Cu-(111) and Cu-(200) peaks, which overlap the fcc Fe peaks. Examination of films deposited on an Al substrate verified that

Table II.  Predominant Phases Detected by X-ray Diffraction

| | |
|---|---|
| Arc-melted 304 | γ + δ |
| Melt spun 304 | γ |
| Sputtered 304 | δ |
| Arc-melted 304-Zr | δ + intermet |
| Melt spun 304-Zr | δ + intermet |
| Sputtered 304-Zr | Amorphous |

the fcc peaks were from the substrate, and it was concluded that the
structure of the sputtered 304 was completely δ-ferrite. The absence of bcc
Fe-(200) diffraction peaks indicated that the deposited film had a preferred
orientation.

The X-ray results for the arc-melted and melt spun 304-Zr exhibited peaks
corresponding to δ-ferrite and an unidentified second phase. Most of the
latter peaks fit with "d-spacings" for $Fe_3Zr$, but there were some anomalies,
which prevented a positive identification. The sputtered 304-Zr exhibited a
broad diffraction maximum centered near the fcc Fe,Cu-(111) peak, which is
characteristic of an amorphous phase (8). A weak bcc Fe-(110) peak was
observed for the 304-Zr deposited on mica.

Mössbauer Spectroscopy

The CEMS spectra recorded at room temperature confirmed the X-ray
diffraction results for the sputtered films. The 304 films exhibited a six-
line spectrum, characteristic of ferromagnetic bcc δ-ferrite. The 304-Zr
films exhibited a broadened single peak centered at -0.14 mm/s (relative to a
pure α-iron calibration standard), which is indicative of an amorphous phase
(Fig. 1). The Mössbauer spectroscopy results are summarized in Table III.

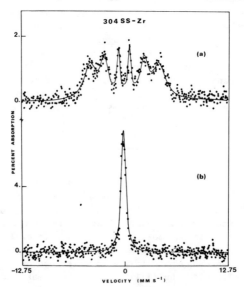

Figure 1 - Conversion electron Mössbauer spectra of sputtered films

(a) 304, (b) 304-Zr

Table III.  Mössbauer Spectroscopy Results

|  | Arc melted | Sputtered | |
|---|---|---|---|
|  | 304 | 304 | 304-Zr |
| Chemical shift (mm/s) | -0.09 | +0.01 | -0.14 |
| Internal magnetic field (kOe) | 0 | 262.6 | 0 |
| Line width (mm/s) | 0.33 | 0.28 | 0.66 |
| Type of SRO around Fe atoms | fcc | bcc | amorph. |

220

## Metallography

The microstructure of the arc-melted 304 was predominantly austenite with a network of ferrite (Fig. 2), which is typical of cast austenitic stainless steel or weld metal. By SEM, the microstructure of the melt-spun 304 appeared to be single phase, but there were two distinct morphologies; either columnar grains aligned at an angle of about 80° to the

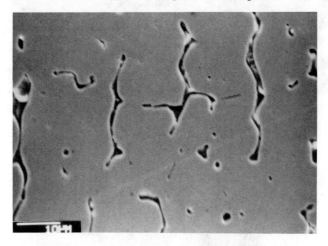

Figure 2 - Arc-melted 304 (modified Murakami etch)

ribbon surface (Fig. 3a), or a cellular morphology (Fig 3b). There was no direct correlation between solidification morphology and casting speed or ribbon thickness. In some cases both morphologies were observed in the same ribbon. Both morphologies exhibited elongated (e.g., Fig. 3a @ A) and equiaxed (e.g., Fig. 3b @ B) substructure features, but the substructure was not well resolved in the SEM micrographs, especially for the higher cooling rates. The limited TEM study clearly showed the columnar fcc grains end-on (Fig. 3c). The substructure of these grains was uniform dislocation tangles with some sub-boundaries. Small (~100 Å) precipitates throughout the γ grains were tentatively identified as molybdenum carbide. Due to the preferential etching of the Cu substrate, the microstructure of the sputtered 304 was not revealed. However, the concentrations of the major alloy elements were determined by energy dispersive X-ray microanalysis (SEM) on the transverse section and wavelength dispersive microanalysis (microprobe) across the film surface (Table IV).

The arc-melted 304-Zr had a 2-phase eutectic-like structure (Fig. 4a). Microanalysis (Table IV) showed that the composition of the second phase was near to $Fe_3Zr$ with substantial concentrations of Cr, Ni, and Mn. The melt-spun 304-Zr contained the same phases, i.e. ferrite and Fe-Zr-Cr-Ni-Mn intermetallic (Table IV), but in this material the intermetallic occurred as discrete particles (Fig. 4b). The SEM-EDS microanalysis was carried out on areas of the intermetallic phase larger than the 2.2 μm spatial resolution at 20 kV accelerating voltage. The etched sections of the sputtered 304-Zr

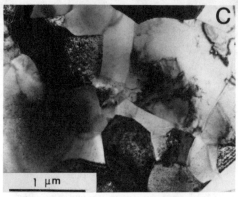

Figure 3 - Melt-spun 304   (a)30 m/s casting speed, 20 μm ribbon thickness;
SEM (chrome regia electrolytic etch)   (b) 30 m/s, 50 μm;SEM (chrome regia
electrolytic etch)   (c) 30 m/s, 20 μm; TEM.

Table IV.  Microanalysis results, at%

|  | Cr | Ni | Mn | Mo | Zr | Fe |
|---|---|---|---|---|---|---|
| Arc-melted 304-Zr* (average) | 16.2 | 8.2 | 1.6 | - | 9.3 | 60.9 |
| Arc-melted 304-Zr* (interm cmpd) | 9.9 | 9.9 | 1.6 | - | 19.0 | 59.0 |
| Melt-spun 304-Zr* (δ matrix) | 18.0 | 7.4 | 1.8 | - | 8.1 | 64.8 |
| Melt-spun 304-Zr* (interm cmpd) | 12.2 | 7.0 | 1.2 | - | 19.6 | 60.0 |
| Sputtered 304** (average) | 17.3 | 8.6 | 1.7 | 0.1 | 0.02 | 72.3 |
| Sputtered 304-Zr** (average) | 16.3 | 6.7 | 1.7 | 0.1 | 10.0 | 66.2 |
| Sputtered 304-Zr*** (upper layer average) | 16.5 | 7.1 | 1.0 | 0.7 | 9.4 | 65.3 |
| Sputtered 304-Zr* (upper layer average) | 16.6 | 9.2 | 1.2 | - | 10.7 | 62.3 |
| Sputtered 304-Zr* (under layer average) | 18.1 | 9.1 | 2.6 | - | 0.03 | 70.2 |

  *  SEM-EDS, transverse section

 **  Microprobe-WDS, film surface

***  SEM-EDS, film surface

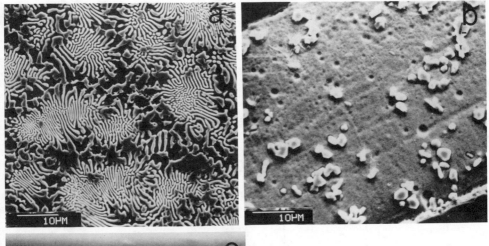

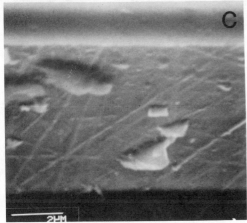

Figure 4 - 304-Zr. (a) arc-melted
(Marble's etch); (b) melt-spun (Marble's
etch); (c) sputtered (10% chromic acid
electropolish)

again did not reveal any microstructure, but they did show two distinct
layers (Fig. 4c). The upper layer (remote from the substrate) contained
10.7 a/o Zr whereas the lower layer contained no zirconium (Table IV).

## Discussion

### 304

For arc melting, the vertical section of the Fe-Cr-Ni phase diagram at
70 wt % Fe (Fig. 5) can be used to explain the as-cast microstucture. The
chromium- and nickel equivalents are given by (9,10):

$$Cr\text{-}eq = \%Cr + \%Mo + 1.5\%Si + 0.5\%Nb$$

$$Ni\text{-}eq = \%Ni + 30(\%C + \%N) + 0.5\%Mn$$

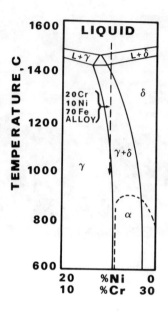

Figure 5 - Vertical section of the
Fe-Cr-Ni phase diagram at 70 wt % Fe (3)

When Cr-eq/Ni-eq > 1.48, the steel solidifies as δ ferrite (11). In the
present case, Cr-eq/Ni-eq = 1.52. Hence, the as-solidified microstructure
would have been primary δ-ferrite dendrites, with some interdendritic γ+δ.
During cooling, most of the δ transforms to γ, resulting in the observed
microstructure of γ grains outlined with a network of δ ferrite (Fig. 2).

The melt-spun 304 ribbon has a similar phase mixture as the arc-melted
material (predominantly γ with some δ), but the microstructure is quite
different. The columnar grains aligned slightly off the perpendicular to the
ribbon surface in the direction of the wheel (Fig. 3a) are characteristic of
partitionless solidification at large undercooling (12). The undercooling for
melt spun Fe-Ni has been estimated previously to be 50-200 K (12), and this
is clearly sufficient to cause the alloy to solidify directly as γ (11,12).
The small grain diameter stabilizes the austenite and precludes
transformation to martensite (1). It is possible that the two observed
solidification microstructures relate to the local cooling rate in the
ribbon. To achieve partitionless solidification, both thermodynamic and
kinetic conditions must be met (13), i.e., the solidification temperature
must be below a critical value, $T_o$, which is roughly half-way between the
solidus and liquidus temperatures, and a critical cooling rate must be
exceeded. The local cooling rate is determined by a number of factors:
ribbon thickness, distance from wheel surface, degree of contact with the
wheel, and the latent heat of solidification (recalescence). Unfortunately,
it was not possible to control all of these variables systematically to
positively demonstrate the effect of cooling rate on solidification
microstructure.

The sputtered 304 films have a completely δ ferrite structure, consistent
with the results of other workers (7,14,15). This suggests that the
effective undercooling during deposition of the sputtered film is much

smaller than for solidification by melt spinning (although the cooling rate is very high for sputtering). Referring to Fig. 5, it can be seen how, under conditions of low undercooling and high cooling rate, 304 could solidify as primary ferrite grains, which remain untransformed to ambient temperature. The sputtering process favours nucleation of the deposit phase (whether equilibrium or metastable) at relatively low undercooling. The substrate is an effective nucleation site, especially when it has been previously sputter cleaned which preferentially etches certain grain orientations. The significance of substrate epitaxy is indicated by the preferred orientation of the sputtered 304 films. It has also been shown that the energy deposition rate at the deposit surface can be such to maintain the temperature ~200 K above that of the substrate (16), and this would further restrict the amount of undercooling. Alternatively, it can simply be argued that bcc ferrite is the equilibrium structure for sputtered 304 with low (<350°C) substrate temperatures (14,17).

## 304-Zr

The arc-melted and melt-spun 304-Zr contain a mixture of $\delta$ ferrite and an intermetallic phase which is based on $Fe_3Zr$ and contains some Cr, Ni and Mn. Thus, Zr acts as a ferrite stabilizer in Fe-Cr-Ni alloys, similar to other Group 4B elements Ti and Hf. The intermetallic phase appears to form by a eutectic reaction when arc melted (Fig. 4a). In contrast to 304, the melt-spun 304-Zr shows evidence of extensive alloy partitioning in the formation of dispersed particles of the intermetallic phase. This suggests that solidification occurs in the latter case at relatively low undercooling, possibly due to nucleation of the intermetallic phase in the melt.

The sputtered 304-Zr films comprise an upper layer which contains 10.0 ± 1.8 a/o Zr, and an underlying layer which is zirconium-free (Fig. 4c and Table IV). Since CEMS samples material to a depth of about 1000 Å, the results pertain exclusively to the ~2 μm-thick upper layer. The conclusion that the upper layer is amorphous is based on the appearance of the broadened single peak centred at -0.14 mm/s (relative to pure $\alpha$-iron), which is similar to the spectrum reported for amorphous $Fe_{90}Zr_{10}$ (8). It was also established by caluclation that the concentration of non-magnetic solute atoms (Cr,Zr) is not high enough to produce the observed line broadening in crystalline iron.

The X-ray diffraction measurements sample the sputtered film to a depth of about 5 μm. Hence, the X-ray results are more difficult to interpret because they eminate from the upper layer and the underlayer. Sputtered 304-Zr film about 10 μm thick was deposited on mica, peeled off the substrate, and X-ray diffraction analysis carried out on both surfaces. The spectrum from the top surface showed the broad diffraction maximum seen in the 304-Zr sputtered on copper and a sharp Fe-(110) peak. By contrast, the spectrum from the surface adjacent to the substrate showed only sharp diffraction maxima.

The microstructures of the sputtered layers could not be revealed, but it appeared that the top layer is much harder and more resistant to etching than the underlayer (Fig. 4c). Finally, by testing with a small magnet, the sputtered 304-Zr films were shown to be magnetic. Since the CEMS results show that the upper layer has zero internal magnetic field (Table III), the underlayer must be the ferromagnetic phase.

The interpretation which best fits all of these observations is that the upper layer is amorphous and the underlayer is crystalline. The effect of Zr in stabilizing the disordered (amorphous) structure in sputtered 304 has been discussed elsewhere (4). An amorphous phase is obtained when the deposit contains at least 10 a/o Zr and sputtering is carried out under conditions of low metal gas thermalization. The glass-forming ability of Zr depends on the atomic misfit of Zr atoms in the Fe lattice.

The "double layer" was present in all the sputtered 304-Zr films examined. This phenomenon could result from a variable rate of Zr deposition or migration of Zr to the surface during the sputtering process. Several mechanisms can be proposed in support of both explanations, but experimental verification is yet lacking. A similar enrichment of boron at the surface of sputtered Fe-Ni-Mo-B has been reported (18). It is important to understand how the parameters of the sputtering process control the production of the "double layer" because it could lead to a technique for depositing 1 μm thick amorphous films.

## Conclusions

1. In arc melting, the solidification sequence follows the near-equilibrium phase relationships. 304 solidifies as bcc ferrite and transforms to fcc austenite during cooling. 304-Zr solidifies as a eutectic comprising ferrite and an intermetallic phase.

2. Zirconium acts as a ferrite stabilizer in Fe-Cr-Ni alloys.

3. Melt-spun 304 solidifies directly as fcc austenite, mainly by planar-front, partitionless solidification at high undercooling. Melt-spun 304-Zr solidifies as a dispersed intermetallic phase in bcc ferrite, with extensive alloy partitioning and low undercooling.

4. Sputtered 304 solidifies as bcc ferrite at low undercooling.

5. Under certain conditions of sputtering, Zr stabilizes an amorphous phase in 304.

## Acknowledgements

The authors are grateful to Peter Rudkowski, Bernard Durocher, Edward Cousineau, and Mark Charest for melt spinning, metallography, X-ray diffraction, and transmission electron microscopy, respectively. The critical comments and suggestions of Dave Embury are much appreciated.

## References

1. B. Cantor, Rapidly Solidified Amorphous and Crystalline Alloys, B.H. Kear, B.C. Giessen and M. Cohen, eds., North Holland, New York, 1982, pp. 317-330.

2. J.V. Wood and J.V. Bee, Chemistry and Physics of Rapidly Solidified Materials, B.J. Berkowitz and R.O. Scattergood, eds., TMS, Warrendale, 1983, pp. 95-107.

3.  L.E. Collins, Can. Met. Quart., vol.25, 1986, pp. 59-72.

4.  Z.S. Wronski, "Role of Zirconium in Forming Thick Amorphous Films of Steel by RF Sputtering", CANMET Report, PMRL 87-36 , 1987.

5.  M. Nose and T. Masumoto, Sci. Rep. Res. Inst. Tohoku Univ., vol.28A Supp., 1980, p.232-241.

6.  I. Inoue, H. Tomioka, and T. Masumoto, J. Mat. Sci., vol.18, 1983, pp.153-160.

7.  T.W. Barbee, B.E. Jacobson, and D.L. Keith, Thin Solid Films vol.63, 1979, pp. 143-150.

8.  Z.S. Wronski, X.Z. Zhou, A.H. Morrish and A.M. Stewart, J. App. Phys., vol 57, No. 1, 1985, pp 3548-3550.

9.  A. Schaeffler, Metal Progress vol.56, no.5, 1949, p.680.

10. W.T. DeLong, Welding Research Supp., July 1974, pp.273s-286s.

11. B.P. Bewlay and B. Cantor, Proceedings of Conf. on Rapidly Solidified Materials, San Diego, 1986, pp.15-21.

12. C. Hayzelden, J.J. Rayment, and B. Cantor, Acta Met., vol.31, 1983, pp.379-386.

13. R.W. Cahn, "Alloys Rapidly Quenched from the Melt", in Physical Metallurgy 3rd ed., R.W. Cahn and P.Haasen eds., Elsevier, 1983, pp.1779-1852.

14. S.D. Dalgren, Met. Trans. vol.1, 1970, pp.3095-3099.

15. P.M. Fabis, J. Vac. Sci. Technol. vol.A5(1), 1987, pp. 75-81.

16. T.W. Barbee and D.L. Kelly, Synthesis and Properties of Metastable Phases, E.S. Machlin and T.M.Rowland, eds., TMS, Warrendale, pp. 93-113.

17. P.J. Grundy and J.M. Marsh, J. Mat. Sci. Let. vol.13, 1978, pp.677-681.

18. M. Rivoire, R. Krishnan, P. Rougier, and J. Sztern, J. App. Phys., vol.52, 1981, pp.1853-1855.

# MACROSEGREGATION IN UNDERCOOLED Pb-Sn EUTECTIC ALLOYS

H.C. de Groh III and V. Laxmanan*

NASA Lewis Research Center, MS 105-1
Cleveland OH. 44135
*Concurrently Dept. of Metallurgy and Materials Sci.,
Case Western Reserve University
Cleveland OH. 44106

## Abstract

A novel technique resulting in large undercoolings in bulk samples (23g) of lead-tin eutectic alloy is described. Samples of eutectic composition were processed with undercoolings ranging from 4 to 20 K and with cooling rates varying between 0.04 to 4 K/sec. The final macrostructure of undercooled samples depends on both the initial undercooling of the melt and the cooling rate. Gravity driven segregation is found to increase with increasing undercooling. A eutectic Pb-Sn alloy undercooled 20 K and cooled at 4 K/sec had a composition of about Pb-72wt% Sn at the top and 55% Sn at the bottom. Macrosegregation in these undercooled lead-tin eutectic alloys is shown to be primarily due to a sink/float mechanism caused by the difference in density of the solid and liquid phases and the undercooling and nucleation behavior of the alloy.

Solidification Processing of Eutectic Alloys
D.M. Stefanescu, G.J. Abbaschian and R.J. Bayuzick
The Metallurgical Society, 1988

# Introduction

The microstructures of important engineering eutectic systems, such as Pb-Sn, Al-Si and cast iron, are strongly influenced by the undercooling achieved prior to solidification, the nucleation characteristics of the alloy, and the cooling rate [1,2]. Gravity-driven macrosegregation in cast irons and other eutectic systems is also well known [3]. Gravity driven macrosegregation is typically caused by differences in liquid density due to gradients in composition and/or differences in density between the liquid and solid phases. In general, the effects of such segregation are detrimental to the engineering properties of the material. The density difference between the liquid and solid phases, and the ability of the Pb-Sn eutectic to be undercooled, are favorable for the study of macrosegregation and the influence of undercooling on segregation behavior. Our objectives are to study how macrosegregation is affected by the initial undercooling of the melt and then to devise means to control it.

The nucleation behavior of Pb-Sn has been studied using fine droplets (0.0004 grams) [4,5,6,7]. In these studies the nucleation behavior was shown to be non-reciprocal; i.e. the Pb-rich phase is a poor nucleant for the Sn-rich phase, while the Sn-rich phase appears to be an effective nucleant for the Pb-rich phase. Thus in a hypoeutectic Pb-rich alloy, the enriched liquid remaining after growth of Pb dendrites may undercool to a temperature below the equilibrium eutectic temperature prior to nucleation of the Sn-rich phase [8,9]. Also, it was shown that the Pb-rich phase is the primary phase nucleating when a eutectic or near eutectic alloy is undercooled. In the eutectic alloy, the heavier Pb-rich dendrites which grow first may settle to the ingot bottom; however, such sedimentation processes, and other macrosegregation phenomena, cannot be studied using fine droplets. Undercooled, larger sized, samples are needed to study these effects.

Most of the bulk undercooling literature has centered around studies of dilute alloys [10], dendrite velocity measurements [11], and alloys with small differences in density between components [12,13]. Davis et al. studied grain structure and solute distribution in undercooled, pure Bi and Bi-100 ppm Ag alloys. They reported no significant macrosegregation; and none should have been expected due to the small difference in composition between the solid and liquid phases. The structure and mechanical properties of undercooled Ni-30% Cu, Ni-20% Cr, and Fe-25% Ni alloys were analyzed in a study by Kattamis and Mehrabian; these authors found the structure and composition of the undercooled alloys to be highly uniform and homogeneous. Macrosegregation is again not expected due to the small variations in densities between phases. The lead-tin system has much larger density differences between phases, compared with the alloys mentioned above. Assuming ideal solutions and 3 vol% shrinkage during solidification, the density ratio of the solid Pb-rich phase of the eutectic (Pb-19.2% Sn) and the eutectic liquid (Pb-61.9% Sn) is about 1.2 . Johnson and Griner observed macrosegregation in a study of Pb-Sn eutectic alloys in which the gravitation level was varied through use of a centrifuge [14].

The objective of this research was to determine how macrosegregation is affected by initial undercooling and the cooling rate prior to nucleation. Bulk sized samples (23g) of Pb-Sn eutectic were solidified with initial undercoolings of between 4 and 20 K and cooling rates ranging from 0.04 to 4 K/sec. Post-solidification metallography of the overall structures was used to help determine whether gravity-driven convection in the liquid state or gravitational settling of dendrite fragments was the primary mechanism leading to macrosegregation. Microprobe analyses of composition variations was conducted to quantitatively determine macrosegregation levels.

## Experimental Procedure

Samples used in this study were prepared from 99.999% pure Pb and Sn with a composition of Pb-61.9wt.% Sn +0.1%. The cleaning and conditioning procedures used for the different samples varied depending on the undercooling desired. All samples were 23g + 1g. The alloys were processed in 1/2 inch diameter borosilicate glass test tubes. The molten alloy was covered with diffusion pump oil to prevent wetting of the crucible and also to promote easy removal of the solidified sample. A small tube furnace, open at the top, was used to permit easy removal of the crucible and sample at the end of the experiment. The processed sample was either quenched in oil or slow cooled in an insulated beaker. All experiments were conducted in a glove box kept at a positive pressure of argon. The temperature of the samples were measured using two quartz sleeved, Type K, thermocouples placed 1.4 cm apart in the metal bath. (See Fig. 1 for approximate locations of TC's for all samples.) The bottom thermocouple was placed 0.5 cm up from the bottom in all experiments. The temperature-time plot of one of the slow cooled ingots (Ingot #3) is shown in Appendix A. Temperatures were recorded with a Hewlett Packard computer (HP-85B) and Data Logger (HP-3497); the data collection rate during cooling was set, depending on the cooling rate, to either once per second or once every five seconds. The accuracy of the thermocouples was periodically checked using pure Sn and found to be within 0.5 K at $232^{o}C$.

Samples in which high undercoolings were desired were thermally cycled. During cycling the samples were mixed with oil (Dow Corning 704). The used oil was decanted after each melting and solidification cycle. The undercooling achieved increased with each cycle and reached its maximum value in the third or fourth cycle. (This process is discussed in more detail elsewhere [9].) Samples in which small undercoolings were desired were not cycled and were solidified with the original oil in which they were stirred. When still less undercooling was desired, Cu powder (0.06mm dia.) was used to nucleate the alloy. The Cu powder was added to the bath during cooling, at a temperature of 220 $^{o}C$. The powder coated the outside of the sample; no Cu particles were found in the bulk of the solidified sample after sectioning. In one sample, a stainless steel screen (50 mesh) was held at the approximate center of the bath, the plane of the screen perpendicular to gravity. This screen experiment helped to clarify certain aspects of the proposed sedimentation process. The procedure used for each experiment is briefly summarized in Table 1.

All samples were cut longitudinally, mounted in metallographic epoxy and polished for optical microscopy. Macrosegregation along the vertical axis of the samples was measured using a microprobe. The average compositions of fifty-one, 0.1mm x 0.1mm areas, spaced evenly between the top and bottom of the samples, were measured using the microprobe. The microprobe was calibrated using pure Pb and Sn standards.

## Results

In Table 1 the undercooling, cooling rate, and thermal gradient just prior to nucleation, and a brief summary of the experimental procedure are presented. A small positive thermal gradient is present from bottom to top, with the top being hotter than the bottom prior to nucleation. Undercooling at the bottom is thus somewhat larger than at the top. The undercoolings listed are those recorded by the bottom thermocouple. The minimum and maximum undercoolings at the extreme top and bottom of the

231

ingot can be estimated by assuming a linear thermal gradient. For example, the ingot with the largest thermal gradient had a estimated undercooling of about 6 K at the very top and 24 K at the very bottom. (The Pb-Sn phase diagram is included in Appendix B.)

Table 1. Thermal Data of Undercooled Pb-61.9% Sn Eutectic Samples

| Description | Undercooling K | Cooling Rate K/sec | Thermal Gradient K/cm |
|---|---|---|---|
| #1 Slow cool, nucleated with Cu powder | 4.2 | 0.066 | 4.4 |
| #2 Slow cooled | 8.0 | 0.040 | 3.2 |
| #3 Slow cooled, thermally cycled | 20.6 | 0.042 | 2.6 |
| #4 Oil quenched, thermally cycled | 20.8 | 4.1 | 6.9 |
| #5 Slow cooled, thermally cycled, solidified w/ s.s. screen[1] | 18.5 | 0.1 | 3.4 |

In Figure 1 through 5 photomicrographs of the longitudinal cross sections and composition measurements of the samples listed in Table 1 are shown. The microprobe measurements were normalized so that the average composition after normalization was equal to the initial alloy composition.[2] The average of the raw, not normalized, microprobe composition measurements are included in Appendix C.

Figure 1 shows the longitudinal cross section of the slow cooled sample (#1) in which Cu powder was used to nucleate the bath and minimize undercooling. A few Pb-rich dendrites are seen at the very bottom of this sample. The rest of the sample is void of dendrites. The microstructure indicates predominately lamellar eutectic and, in some areas, non-lamellar eutectic. Grains of eutectic can be seen growing vertically up, directionally, in the top 2/3 of the sample. The segregation in this sample, undercooled 4.2 K, is limited to local variations, as shown by the

[1]    Only one thermocouple was used in this experiment; the thermal gradient was estimated to be that of the average of the other slow cooled samples. Since the thermocouple which measured the undercooling (18.5 K) was 1.5 cm above the bottom, the maximum undercooling in this sample is probably closer to 23 K.

[2]    The data were normalized by multiplying each value by the ratio of the actual, as weighed composition of the alloy to the average of the measured data, that is: Normalized datum=measured datum x(actual ave.comp./measured ave.comp.)

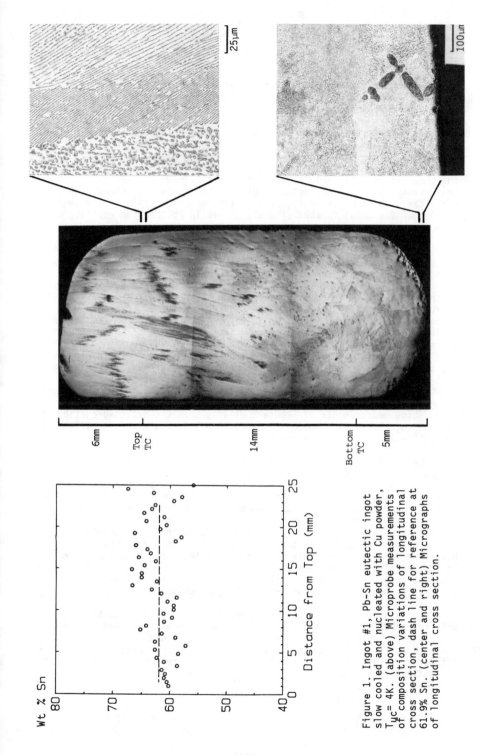

Figure 1. Ingot #1, Pb-Sn eutectic ingot slow cooled and nucleated with Cu powder, $T_{uc}$= 4K. (above) Microprobe measurements of composition variations of longitudinal cross section, dash line for reference at 61.9% Sn. (center and right) Micrographs of longitudinal cross section.

composition measurement in Figure 1. The average value is very nearly the starting composition, indicating no significant macrosegregation. The Pb-rich dendrites at the sample bottom do not significantly affect the average composition.

The structure and composition variation of ingot #2 are shown in Figure 2. Spherical, cylindrical, and cross (+) shaped Pb-rich dendritic particles, surrounded by eutectic, are seen randomly arranged in the bottom of the ingot. No Pb-rich dendrites are present in the top half of the sample. The eutectic grains seen above the Pb-rich dendrites are believed to have grown upwards (after nucleation and growth of the Pb-rich dendrites) in the direction of the thermal gradient which remained after recalescence. Again, there is no clear indication of macrosegregation from the composition measurements shown in Figure 2. However, the six measurements taken in the bottom region, which contained the Pb-rich dendrites, indicated an average composition of Pb-58 wt% Sn. The lower portion of the ingot is thus slightly richer in Pb than the equilibrium eutectic.

Figure 3 shows the structure and composition measurements of ingot #3, the slow cooled ingot which was thermally cycled and undercooled 20.6 K prior to nucleation. The top 1/3 of this ingot consists of grains of primarily ordered groups of Sn-rich dendrites and interdendritic eutectic. In the approximate center of the ingot are columnar grains of eutectic free of Pb and Sn-rich dendrites. On the right-hand side of the ingot a eutectic solidification front can be seen which grew up from the bottom region. Eutectic grains also grew down from the grains of Sn-rich dendrites above, to form a eutectic grain boundary at which the two fronts met. It is also apparent that the right-hand side of the ingot, near this eutectic region, was pulled away from the tube wall due to shrinkage. This shows that the dendrite free eutectic areas in the approximate center of the ingot were last to solidify. In the bottom, approximate 1/2 of the sample, are Pb-rich dendrites with a considerable amount of eutectic. The composition measurements of Figure 3 clearly show the macrosegregation. The average composition at the top is about Pb-68% Sn while the composition at the bottom is approximately Pb-58% Sn.

The oil quenched and thermally cycled sample, ingot #4, is shown in Figure 4. This sample was undercooled about the same amount as ingot #3, however, the cooling rate just prior to nucleation was about 100 times faster. In the bottom 1/3 of the ingot are the round, rod and cross shaped Pb-rich dendrite fragments characteristic of the other samples. The volume fraction of Pb-rich dendrites to eutectic is greatest at the very bottom and decreases toward the center. The top 1/3 of the ingot consists of grains of Sn-rich dendrites and eutectic. In the center of the ingot are blossoming grains of eutectic which resemble those of an equiaxed zone. The composition measurements shown in Figure 4 indicate this sample to be the most severely segregated. The average compositions at the top and bottom are about Pb-72% Sn and Pb 55% Sn respectively.

Figure 5 shows the structure of ingot #5, the Pb-Sn eutectic ingot, undercooled 20 K, and solidified with a stainless steel screen held near the center. The macrosegregation process in this sample is very similar to that proposed for the two previously discussed ingots, only the screen has not permitted the Pb-rich dendrites which formed near the top of the sample to settle to the bottom.

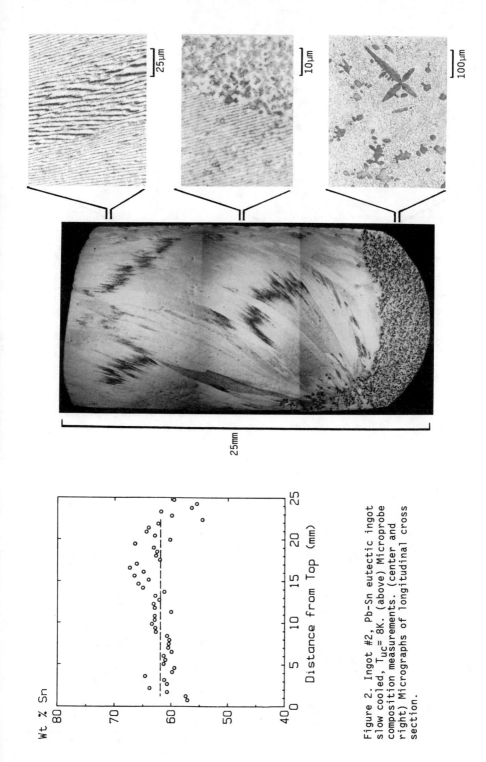

Figure 2. Ingot #2, Pb-Sn eutectic ingot slow cooled, $T_{uc}$= 8K. (above) Microprobe composition measurements. (center and right) Micrographs of longitudinal cross section.

235

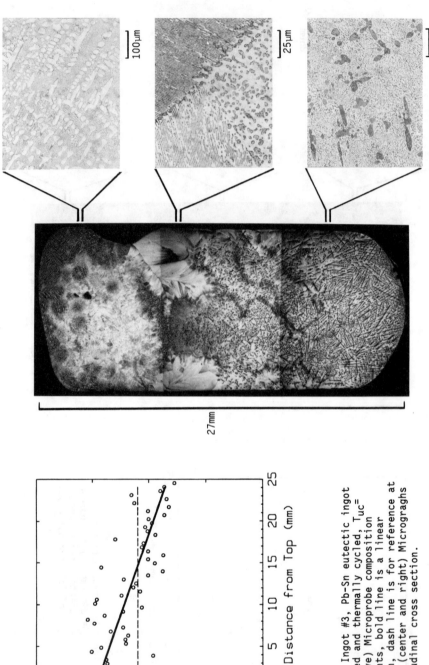

Figure 3. Ingot #3, Pb-Sn eutectic ingot slow cooled and thermally cycled, $T_{uc}$= 20K. (above) Microprobe composition measurements, bold line is a linear regression, dash line is for reference at 61.9% Sn. (center and right) Micrograghs of longitudinal cross section.

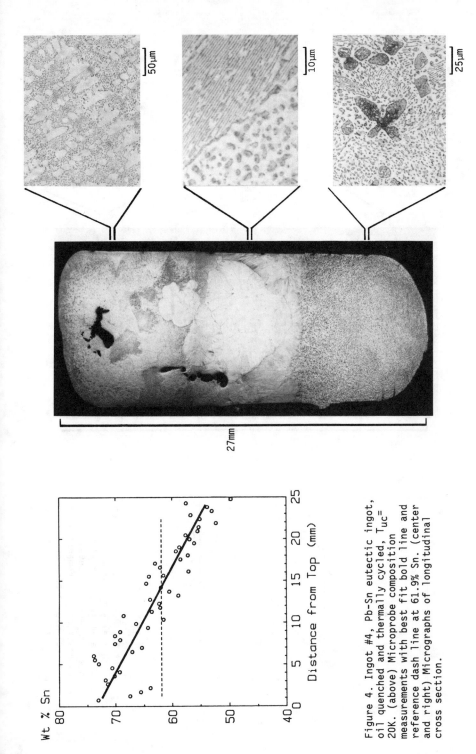

Figure 4. Ingot #4, Pb-Sn eutectic ingot, oil quenched and thermally cycled, $T_{uc}$= 20K. (above) Microprobe composition measurements with best fit bold line and reference dash line at 61.9% Sn. (center and right) Micrographs of longitudinal cross section.

237

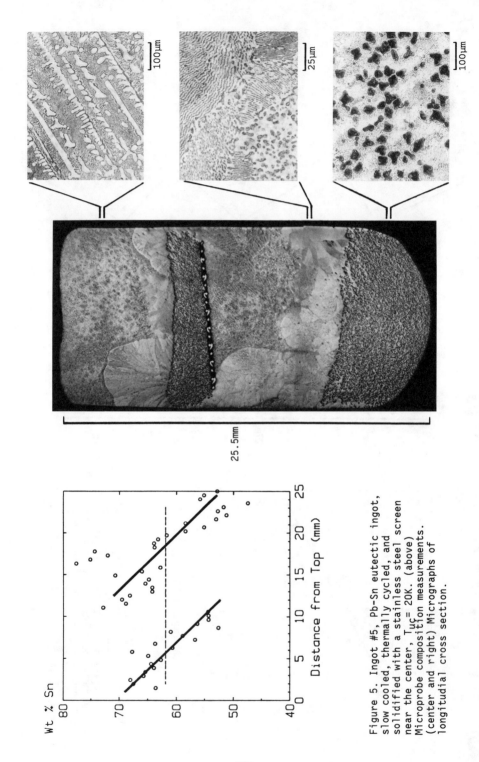

Figure 5. Ingot #5, Pb-Sn eutectic ingot, slow cooled, thermally cycled, and solidified with a stainless steel screen near the center, $T_{uc}= 20K$. (above) Microprobe composition measurements. (center and right) Micrographs of longitudial cross section.

## Discussion

Both in this study and in related work [9] the Pb-rich phase was found to be the primary phase growing in an undercooled Pb-Sn eutectic alloy. The extent of (Pb-rich) dendritic solidification depends on the amount of undercooling prior to the first nucleation event; the fraction of Pb-rich dendrites formed is higher at higher undercoolings. With higher undercoolings and more Pb-rich dendrites, the liquid becomes more rich in Sn. With liquid compositions more rich in Sn at higher undercoolings, extensive remelting of the Pb-rich dendrites appears to take place.

Our experimental observations may be explained on the basis of the above hypothesis. Thus, in ingot #1, which had a small undercooling, only a small amount of Pb-rich dendritic solidification took place and no significant macrosegregation. The composition of the solidifying two phase solid is nearly the same as the liquid. The maximum undercooling in ingot #1, about 6 K, was just enough to drive the small amount of metastable Pb-rich dendritic solidification observed.

A greater fraction of Pb-rich dendrites are seen with undercoolings of 8 K or more, as in ingots #2 through #5. At these undercoolings dendrite tip velocities, estimated from earlier measurements by others, are of the order of at least a few centimeters per second [15,16]. This would allow the rapid growth of a dendritic skeleton across the highly undercooled portion of the ingot. However, dendritic remelting, occurring during and immediately following recalescence, fragments this skeleton which breaks into spherical, cylindrical, and cross (+) shaped dendritic particles [13,17]. These fragments then settle to the bottom, as seen in Figure 2, causing macrosegregation.

No indication of buoyancy driven convection in the liquid state was observed in this study. It is likely in ingot #2 that the Pb-rich dendrites grew only in the bottom half of the ingot, the portion of the ingot undercooled at least 5 K. This alone would not result in any macrosegregation. For macrosegregation to develop, movement of the now Sn-rich liquid up and/or the Pb-rich dendrites down is required. Since the density difference between the solid and liquid phase is much greater than those in the liquid due to composition and temperature gradients in the liquid, the settling of the solid dominates. This is shown in the composition variation of ingot #2, Figure 2. The Pb-rich dendrites which grew in the bottom half of the ingot sank to the bottom fifth, leaving the liquid above richer in Sn as shown.

A number of experiments were done to further test for convection and segregation in the liquid. The mid-ingot screen of #5 was designed to stop solid sedimentation and not significantly impede segregation in the liquid. During the time the bath is undercooled (about 3 min. in the slowly cooled ingots) it is possible that clusters of Pb and/or Sn atoms form and sink and float respectively. If this were to occur to a significant extent a channel of Pb-rich and/or Sn-rich material would be seen at the edges of the screen; this is not the case as seen in Figure 5. Ingot #5 also shows that Pb-rich dendrites can grow throughout the bath and that the segregation is rather insensitive to the small gradients of these experiments. To prevent any segregation in the liquid, the liquid was vigorously stirred prior to cooling. In another experiment, during the final stages of cooling below the eutectic temperature, a small amount of liquid was removed from the top using a small ladle. This liquid was chemically analyzed and found to be of eutectic composition, thus indicating no segregation in the liquid prior to nucleation.

239

With initial undercoolings of about 20 K the volume fraction of Pb-rich dendrites is greater (as compared to ingots solidified with less initial undercooling), thus the composition of the liquid they leave behind after settling is richer in Sn. In the top region of the ingot the liquid is rich enough in Sn to grow Sn-rich dendrites as seen in Figure 3. The increase in undercooling from 8 to 20 K increased the amount of macrosegregation.

At faster cooling rates, such as that used to solidify ingot #4, it is believed that a larger amount of Pb-rich dendrites are able to form during recalescence. The remelt and breakup of the Pb-rich dendrites is thus more extensive because local Sn concentrations are higher. This results in finer Pb-rich dendrite fragments and a greater packing density at the ingot bottom, and the increase in macrosegregation as compared to ingot #3 (Figures 3 and 4).

## Conclusions

1) The amount of macrosegregation in the Pb-Sn eutectic ingot undercooled 4 K was negligible.

2) In ingots undercooled 8 K or more, a significant amount of Pb-rich dendrites are formed. It is suggested that these dendrites grow rapidly into the bath during recalescence, breakup due to remelting, and settle to the bottom causing the macrosegregation observed in these ingots.

3) The extent of (Pb-rich) dendritic solidification increases with increasing undercooling. As Pb-rich dendrites grow, breakup, and settle to the bottom, the remaining liquid in the upper region of the ingot becomes more rich in Sn. With an undercooling of 20 K, the liquid at the top region of the ingot is rich enough to grow Sn-rich dendrites. These effects cause greater amounts of macrosegregation as compared to ingots solidified with less undercooling at the same cooling rate. A Pb-Sn eutectic ingot undercooled 20 K with a cooling rate of 4 K/sec just prior to nucleation had a composition of about Pb-72wt% Sn at the top and 55% Sn at the bottom.

4) No channeling or other indications of segregation in the liquid state, due to density gradients in the liquid or clustering of Pb and/or Sn-rich atoms during undercooling, was observed in this study.

## Acknowledgments

The authors wish to thank the staff of the Microgravity Materials Science Laboratory and others for their contributions and assistance to this project: Richard Rauser for help with metallography, David Lee for help with the experimental work and data collection, Steve White and Bruce Rosenthal for assistance with hardware, Judy Auping for guiding us through the computer software, and Francis Terepka and David Hull for doing the extensive microscopy and microprobe analysis.

240

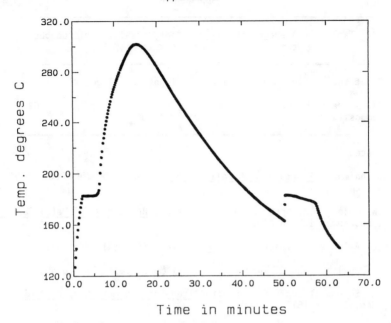

Figure 6. Temperature-Time data measured by the bottom TC of Ingot #3. Data collection rate was once every 10 seconds, then increased to once every 5 seconds during the final stage of cooling. The temperature of the plateau during melting was 182.6 °C. The thermal gradient and cooling rates were measured just before recalescence, about 50 minutes into the experiment.

Appendix B

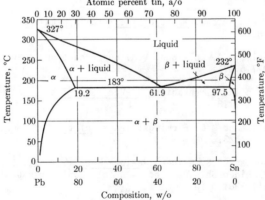

Figure 7. The Pb-Sn phase diagram (ASM Handbook of Metals)

Table 2.   Averages of the raw, not normalized microprobe
composition measurements in wt.% Sn.

| Ingot No. | Ingot #1 | Ingot #2 | Ingot #3 | Ingot #4 | Ingot #5 |
|---|---|---|---|---|---|
| Raw average composition | 63.4 | 64.0 | 68.0 | 63.4 | 69.1 |

References

1) B. Chalmers, Principles of Solidification, R.E. Kriefer Pub., 1964
p.213, 224.

2) R.W. Heine, C.R.Loper and B.C. Rosenthal, Principles of Metal Casting,
1955, reprinted 1980 p.298, 491.

3) B.C. Allen and S.Isserow, Acta Met., Vol. 5, Aug 1957 p.465.

4) J.H. Hollomon and D. Turnbull, Jol. of Met., Sept. 1951 p.803.

5) B.E. Sundquist and L.F. Mondolfo, Trans of the Met. Soc. of AIME, Vol.
221, Feb. 1961 p.157.

6) J.H. Perepezko and J.S. Paik, Proc. of the MRS Symp., Rap. Sol.
Amorphous and Cryst. Alloys. 1982.

7) M.G. Chu, Y. Shiohara and M.C. Flemings, Met. Trans A, Vol. 15 A, July
1984 p.1303.

8) W.V. Youdelis and S.P. Iyer, Met. Sci., TMS, Vol. 9 1975 p.289.

9) H.C. de Groh and V. Laxmanan, pending pub. in Met. Trans. with Sec. Int.
Symp. on Micro. Grav. Mat. Sci. Research, 1988.

10) K.G. Davis and Fryzuk, Trans of the Metallurgical Soc. of AIME, Vol.
233, 1965 p.1983.

11) G.L.F. Powell, G.A. Colligan, V.A. Surprenant and A. Urquhart, Met.
Trans., 8A, 1977 p.971.

12) T.Z. Kattamis and R.Mehrabian, J. Vac. Sci. Tech., Vol. 11 No.6,
Nov./Dec. 1974, p.1118.

13) T.Z. Kattamis and M.C. Flemings, Trans. of the Met. Soc. of AIME, Vol.
236, Nov. 1966 p.1523.

14) M.H. Johnson and C.S. Griner, Scripta Met., Vol. 11, 1977 p.253. 15) K.
Kobyashi and P. Hideo Shingu, Proc. 4th Int. Conf. on Rapidly Quenched
Metals, (Sendai) 1981 p.103.

16) G.T. Orrok, Thesis, Harvard University, 1958.

17) M.C. Flemings, Y. Shiohara, Y. Wu and T.J. Piccone, Undercooled Alloy
Phases, Proc. 1986 Hume-Rothery Symp., Ed. Collings and Koch p.329-341.

Directed-Energy Electron-Beam Processing of a

Hypoeutectic $Cr_{90}Ta_{10}$ Alloy

J. S. Huang and E. N. Kaufmann

Lawrence Livermore National Laboratory
P.O. Box 808, Livermore, CA 94550

## Abstract

A hypoeutectic $Cr_{90}Ta_{10}$ alloy was processed using a directed-energy electron-beam surface melting and resolidification technique to study its microstructure evolution during rapid solidification. The power of the electron-beam was 2500 W and the scan speed ranged from 0.13 to 2.0 m/sec. Microstructure characteristics such as transitions from planar front to dendritic growth, and from cellular to dendritic growth were observed. At low solidification rate the interdendritic regions are characterized by the $Cr/Cr_2Ta$ eutectic, and at high solidification rate the intercellular regions are characterized by a $Cr_2Ta$ phase. For the latter, the distribution of Ta-solute across the interior of a cell is very uniform. In a given sample, the primary cell spacing increases as the solidification front moves from the substrate/regrowth interface toward the surface. The solidification parameters, i.e., temperature gradient and growth velocity, were determined with finite-element heat flow analyses. The observed microstructure characteristics were correlated to these parameters using available theoretical models.

Solidification Processing of Eutectic Alloys
D.M. Stefanescu, G.J. Abbaschian and R.J. Bayuzick
The Metallurgical Society, 1988

## Introduction

Rapid solidification has developed into an important technology for producing novel microstructures of materials. The microstructural morphology of rapidly solidified materials depends upon solidification parameters, such as growth velocity, cooling rate, temperature gradient, etc. Despite a significant amount of past work, relatively few experimental studies have made quantitative correlations between microstructural morphology and solidification parameters during rapid solidification. The directed-energy electron-beamsurface melting and resolidification technique is very useful for such a study. In this technique, a stream of electrons are focused on and scanned across the surface of a material. This creates a melted track which subsequently solidifies. By controlling the total power and scan speed of the electron-beam, solidification parameters can be systematically varied. These parameters can also be estimated by heat transfer analysis and correlated to microstructure. In this paper, we report the results of a study using this technique on the rapid solidification of a near-eutectic composition alloy, $Cr_{90}Ta_{10}$. This study is an extension of a previous study on this alloy which used electron beam and pulsed laser-beam irradiation (1). It was shown there that the irradiated surfaces consisted of: (1), an extended solid-solubility BCC phase under the electron-beam irradiation and (2), a mixture of metallic glass and a FCC phase under the laser irradiation. In this work, we emphasize the regime where the solidification rate is lower, so that cellular microstructures can be observed. Of particular interest is the correlation of cell spacing to solidification parameters. Models for this correlation have been developed by Hunt (2), and by Kurz and Fisher (3). However little work has been attempted to investigate their applicability in the regime of rapid solidification.

## Experimental Details

The alloy was prepared by vacuum arc melting. Disk samples, 4 mm thick, surface polished to a final finish of 1 µm, were used for electron-beam surface processing. The processing was conducted with a Hamilton Standard model W2-OS electron-beam welder. The acceleration voltage of the electron beam was 100 kV, and the current was 25 mA. Scan speeds employed were 0.13, 0.25, 0.5, 1.0, and 2.0 m/sec. All scans were in single-pass (i.e., nonoverlapping) mode and produced melted and resolidified trails about 1.2 - 1.5 mm wide and 0.01 - 0.6 mm deep. The processed samples were cross-sectioned transverse to the trails for metallographic examination. A 10%-oxalic acid etchant was used prior to examination by both an optical microscope and a Hitachi-S800 scanning electron microscope operating at 20 kV. Selected cross-section transmission electron microscopy (TEM) samples were also prepared to examine the crystal structure and compositions of second phases.

## Estimation of Solidification Parameters

Solidification parameters were determined using a two-dimensional finite-element heat transfer analysis code (4), TOPAZ2D, developed at the Lawrence Livermore National Laboratory. Details of the algorithm of this code are documented in reference (4). The two dimensions used in the analysis are the direction of beam scan and the normal to the surface of the sample. The heat flux transverse to the melt trail is neglected. This is a reasonable approximation for our samples where the melt depths are shallow and the melt and solidification fronts are very flat at the bottom of melt zones. The electron-beam irradiation was modeled as a heat-flux boundary condition which varies with position and time. It was assumed that the incident electron-beam had a Gaussian intensity distribution with a first moment (R) of 0.35 mm, implying a melt trail width of 4R. The fraction of back-scattered electrons, which do not contribute to heating, was estimated as 0.23 using the relation given by

Klein (5), which correlates the back-scattered-electron efficiency with the atomic number of the material.

The following material parameters were used for this analysis:

$Cr_{90}Ta_{10}$(4 mm thick)

> specific heat = 1.0 J/g-K
> thermal conductivity = 0.5 W/cm-K
> density = 8.2 g/cm$^3$
> melting point = 2033 K
> latent heat = 283.0 J/g

Cu Substrate(6 mm thick)

> specific heat = .46 J/g-K
> thermal conductivity = 3.34 W/cm-K
> density = 8.96 g/cm$^3$

where for the $Cr_{90}Ta_{10}$ alloy, the latent heat is taken as that of Cr, the specific heat as that of a Cr-23%Fe alloy (6), and the thermal conductivity was obtained by iteration until the calculated maximum melt depth matched the experimental data. The output of the TOPAZ2D code includes time and position dependent temperature distribution, from which the maximum melt depth and the melting-point-isotherm contour are determined. A comparison between the calculated maximum depth and the actual maximum depth is shown in Table I. The difference between them is less than 14%. It was assumed that the solidification direction is perpendicular to the melting-point-isotherm contour, and the temperature gradient in the liquid was calculated along this direction. The solidification velocity was calculated as the electron-beam scan speed multiplied by the cosine of the angle between the beam-scan direction and the solidification direction. The calculated solidification velocity and temperature gradient in the liquid for all the samples are shown in Figs. 1 and 2, respectively.

Table I   Comparison of the measured maximum melt depth with the calculated maximum melt depth for the electron-beam surface-processed $Cr_{90}Ta_{10}$ alloy (power = 2500 W)

| Scan speed (m/sec) | Measured maximum melt depth (μm) | Calculated Maximum melt depth (μm) |
|---|---|---|
| 0.13 | 600 | 556 |
| 0.25 | 310 | 270 |
| 0.5 | 140 | 126 |
| 1.0 | 56 | 54 |
| 2.0 | 15 | 13 |

## Experimental Results

Typical scanning electron photomicrographs for the sample processed at a scan-speed of 0.13 m/sec are shown in Figs. 3a and 3b. The solidification initially has a dendritic morphorlogy with BCC Cr phase as the major phase. The composition of this alloy is slightly below the eutectic composition, therefore as the temperature decreases from the superheated liquid, the first solid which solidifies will be the Cr phase. However as the isotherm velocity increases, the liquid-solid interface undercooling increases and the growth

245

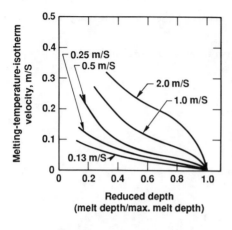

Fig. 1 Calculated solidification velocity versus the position of liquid/solid interface.

Fig. 2 Calculated temperature gradient in liquid versus the position of liquid/solid interface.

is replaced with coupled eutectic growth (Cr and $Cr_2Ta$ phases) (Fig. 3a).
As the solidification proceeds, the growth velocity increases gradually and
eventually the coupled eutectic growth gives way to a dendritic growth with
interdendritic eutectic (Fig. 3b).

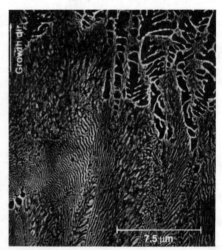

Fig. 3 Microstructures of resolidified $Cr_{90}Ta_{10}$ alloy after being irradiated
with the electron-beam at a scan speed of 0.13 m/sec.: (a) near the
substrate/regrowth interface; (b) at 0.94 $d_{max}$ (maximum melt depth).

Typical scanning electron photomicrographs for the sample processed at a
scan-speed of 0.25 m/sec are shown in Figs. 4a and 4b. The solidification
starts (Fig. 4a) with a dendritic growth with interdendritic eutectic. This
mode of solidification proceeds continuously as the growth front approaches
the surface. As the growth front advances, the size of the dendritic cells
increases (Fig. 4b).

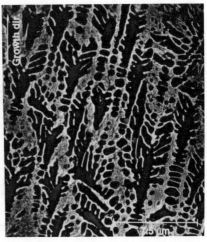

Fig. 4 Microstructures of resolidified $Cr_{90}Ta_{10}$ alloy after being irradiated
with the electron-beam at a scan speed of 0.25 m/sec.: (a) near the
substrate/regrowth interface; (b) at 0.74 $d_{max}$.

Typical scanning electron photomicrographs for the sample processed at
0.5 m/sec are shown in Figs. 5a, and 5b. The solidification now starts with
a cellular or columnar growth morphology (Fig. 5a). However as the growth
front moves forward, it eventually gives way to a dendritic growth (Fig. 5b),
which has a higher tendency for side-arm branching.

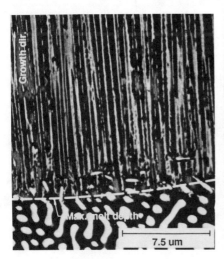

Fig. 5 Microstructures of resolidified $Cr_{90}Ta_{10}$ alloy after being irradiated
with the electron-beam at a scan speed of 0.5 m/sec.: (a) near the
substrate/regrowth interface; (b) at 0.3 $d_{max}$.

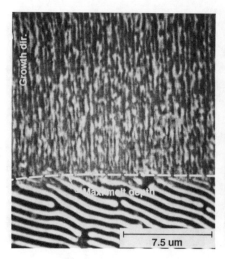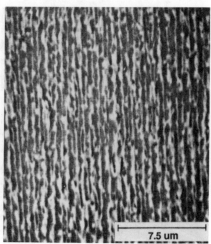

Fig. 6 Microstructures of resolidified $Cr_{90}Ta_{10}$ alloy after being irradiated with electron-beam at a scan speed of 1.0 m/sec.: (a) near the substrate/regrowth interface; (b) at 0.5 $d_{max}$.

Typical scanning electron photomicrographs for samples processed at 1.0 and 2.0 m/sec are shown in Figs. 6 and 7, respectively. For these two samples, the solidification proceeds with a cellular or columnar growth from the substrate/regrowth interface all the way to the surface. There is a strong indication that the cell size increases as the solidification front moves forward (comparing Fig. 6a with 6b). These characteristics will be discussed in more detail in the next section.

Fig. Microstructure of resolidified $Cr_{90}Ta_{10}$ alloy after being irradiated with the electron-beam at a scan speed of 2.0 m/sec.

The interdendritic phase at the lower solidification rates (Figs. 3 and 4) is clearly a $Cr/Cr_2Ta$ eutectic. However, at higher solidification rates, the intercellular phases are narrower and replaced with a Ta-rich phase.

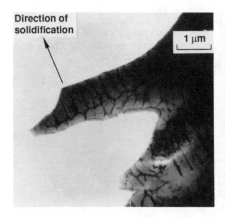

Fig. 8 (a) Transmission electron photomicrograph of the microstructure of the
sample processed at a scan speed of 2.0 m/sec.; (b) converging micro-
beam diffraction pattern from the intercellular region in (a).

Figures 8a and 8b show a cross section TEM photomicrograph of the resolidified microstructure of the sample processed at 2.0 m/sec, and an associated
converging beam diffraction pattern obtained from the intercellular phase.
The photograph is taken from an area at about one-third of the maximum melt
depth, and the diffraction pattern is identified as that of the high
temperature hexagonal Laves phase, $Cr_2Ta$. The Ta solute distribution across
a cell, analyzed by X-ray microanalysis, is shown in Fig. 9. It is interesting
that the distribution of solute across a cell is very uniform. This result
is parallel to an observation in another rapidly solidified hypoeutectic Ag-15
wt.% Cu alloy (7). The apparent content of tantalum at both the cell boundaries is lower than that expected for the $Cr_2Ta$ phase, which is equal to
26.5 atomic percent at the substrate. This is expected since the resolution
limit of the X-ray microanalysis is estimated about 800 Å, which is significantly larger than the width of the second phase at the cell boundaries, about
500 Å.

## Discussion

It is well known that cellular and dendritic solidification in alloys
arise from constitutional supercooling. This occurs, for a given growth
velocity, when the actual temperature gradient is smaller than the liquidus
gradient in liquid in front of a solidification interface. A solidification parameter, G/V, where G is the temperature gradient in the liquid and V
is the solidification velocity, is usually used to describe this kind of
supercooling. For the supercooling not to occur, i.e. planar growth, the
value of G/V has to be high enough for a material. The calculated G/V values
for our case are shown in Fig. 10. During the initial stage of growth near
the substrate/regrowth interface, G/V is the highest for the sample processed
at 0.13 m/sec. This could explain why there is a planar growth of eutectic
for this sample at the beginning, while for other samples, the G/V values are
low enough that cellular and dendritic growth occur. For all the samples,
G/V decreases as the solidification front moves toward the surface, therefore
causing the instability of the solidification front (8), and the transition

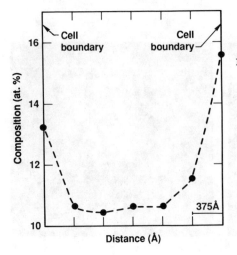

Fig. 9 Distribution of Ta-solute across a cell for the sample shown in Fig. 8.

from eutectic growth to dendritic growth (Fig. 3), as well as the transition from cellular growth to dendritic growth (Fig. 5). This interpretation of data is rather qualitative. As mentioned by Flemings (8), the entire topic of cellular to dendritic transition still remains a good one for theoretical study. It is interesting to note that for samples processed at lower

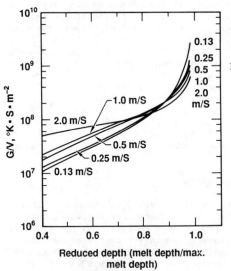

Fig. 10 A plot of temperature gradient to growth rate ratio versus the position of liquid/solid interface.

electron-beam scan speed, the decreasing rate of G/V versus distance is higher (Fig. 10), and dendritic growth is preferred in the solidification.

Fine intercellular eutectic is not seen in the microstructures solidified at high growth rates because eutectic growth is a slow process and seldom exceeds 0.1 m/sec (9). For the microstructure shown in Fig. 8a, the growth velocity is estimated above 0.3 m/sec, therefore eutectics do not form.

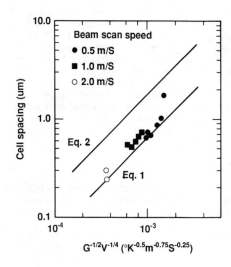

Fig. 11 Calculated primary cell
spacing and measured data
plotted versus $G^{-0.5}V^{-0.25}$.

(Figure axes: Cell spacing (um) vs $G^{-1/2}V^{-1/4}$ ($^{\circ}K^{-0.5}m^{-0.75}S^{-0.25}$); Beam scan speed: ● 0.5 m/S, ■ 1.0 m/S, ○ 2.0 m/S; labeled Eq. 1 and Eq. 2)

One useful quantity used to describe dendritic or cellular structures in
columnar growth is the primary cell spacing. Two detailed theoretical models
have been proposed in the literature (2,3) to characterize this primary cell
spacing as a function of growth rate, temperature gradient and alloy composition. One of our objectives in this work is to investigate whether these
models apply to the rapid solidification regime. The first model was
presented by Hunt (2). It predicts that when growth rate $\gg V_{cs}$, where $V_{cs}$
is the critical velocity above which constitutional supercooling occurs, the
primary cell spacing can be described as

$$\lambda_1 = 2.83(k\Delta T_0 D\gamma/\Delta S)^{0.25} \, V^{-0.25} \, G^{-0.5} \tag{1}$$

where D is the solute diffusion coefficient in the liquid, $\gamma$ is the liquid-
solid interfacial energy, $\Delta S$ is the entropy of fusion per unit volume, k
is the solute partition coefficient, and $\Delta T_0$ is the temperature difference
between the equilibrium liquidus and solidus at the initial solute content of
the alloy, $C_0$.

The second model by Kurz and Fisher (3) predicts that, for growth rate
higher than a transition value defined as $V_{cs}/k$, the primary cell spacing
can be described as,

$$\lambda_1 = 4.3(\Delta T')^{0.5}[\gamma D/(Sk\Delta T_0)]^{0.25} \, V^{-0.25} \, G^{-0.5} \tag{2}$$

where $\Delta T'$ is the temperature difference between the tip and the base of a
cell and the rest of the parameters have the same meanings as above.
Approximating $\Delta T' = \Delta T_0$, Eqs. 1 and 2 differ from each other by a constant
and predict the same relations between the primary cell spacing and temperature
gradient and growth rate. For the solidification we are considering here, the
growth rates are, in general, higher than the transition velocity as described
above, therefore Eqs. 1 and 2 can be compared with our data. An investigation
of the microstructures of the samples processed at 0.5, 1.0 and 2.0 m/sec
(Figs. 6, 7 and 8), indicates that the primary cell spacing increases as the
solidification fronts moves toward the surface. This phenomenon can be
explained using Eqs. 1 and 2. As the solidification front moves toward the
surface, the growth rate increases (Fig. 1), and the temperature gradient

251

(Fig. 2) decreases. Since the $-1/2$ power temperature-gradient dependence of primary cell spacing dominates the $-1/4$ power growth-rate dependence, it can be expected that the primary cell spacing increases when the solidification front moves toward the surface. The measured primary cell spacing from the samples processed at 0.5, 1.0 and 2.0 m/sec are plotted in Figure 11 versus the term, $G^{-0.5} V^{-0.25}$. Shown for comparison are also the values predicted by Eqs. 1 and 2, using the following assumed material parameters: $D = 10^{-9}$ $m^2sec^{-1}$; $\gamma = 0.4$ J $m^{-2}$; $\Delta S = 1.2 \ast 10^6$ $Jm^{-3}$; $k = 0.3$; and $\Delta T' = \Delta T_0 = 24$ K. Due to the imprecise values of these parameters and of the thermal properties used for the analyses of G and V, it is not known how good these comparisons are, however the models do adequately predict the functional relationship between primary cell spacing and $G^{-0.5} V^{-0.25}$.

## Conclusions

We have studied the microstructure characteristics of a rapidly solidified $Cr_{90}Ta_{10}$ hypoeutectic alloy using the directed-energy electron-beam surface melting and resolidification technique. The following can be concluded:

- Cellular or dendritic structures are observed throughout except for a small initial portion of the sample processed at 0.13 m/sec, where a plane front eutectic growth occurs.

- The intercellular or interdendritic phases range from eutectics at low solidification rates to a hexagonal Laves phase, $Cr_2Ta$, at high solidification rates.

- For the sample processed at the highest solidification rate, the distribution of Ta solute across a cell is very uniform and exceeds the maximum equilibrium solubility by about 2.5 times.

- The cell spacing increases as the solidification front moves toward the surface due to decreasing temperature gradient in the liquid. The synergistic effects of temperature gradient and growth velocity on cell spacing are described well with the available theoretical models.

## Acknowledgments

The authors appreciate the help of B. Olsen on processing of the samples, and the help of M. Wall and J. Yoshiyama on the characterization of microstructures using scanning and transmission electron microscopy. The assistance of A. Shapiro, author of TOPAZ2D code, on the numerical analyses is also appreciated. Work performed under the auspices of the U. S. Department of Energy, office of Basic Energy Sciences, Div. of Materials Science, by the Lawrence Livermore National Laboratory under contract No. W-7405-ENG-48.

## References

1. D. A. Alden et al., "Electron Beam and Pulsed Laser Irradiation of Glass-Forming Cr-Ta and Cr-Zr-Ge Alloys", Script Metallurgica, 19 (1985), 67-72.

2. J. D. Hunt, Solidification and Casting of Metals (Metals Society, London, 1979) 1.

3. W. Kurz and D. J. Fisher, "Dendritic Growth and the Limit of Stability: Tip Radius and Spacing," Acta Met., 29 (1981), 11-20.

4. A. B. Shapiro, "TOPAZ2D - A Two Dimentional Finite Element Code for Heat Transfer Analysis, Electrostatic, and Magnetostatic Problems," Lawrence Livermore National Laboratory, UCID-20824, July 1986.

5. C. A. Klein, "Further Remarks on Electron Beam Pumping of Laser Materials," Applied Optics, 5 (12) (1966), 1922-1924.

6. Y.S. Touloukian, and E. H. Buyco, Thermophysical Properties of Metals and Alloys, Vol. 4 (IFI Plenum, New York, 1970), p308.

7. L. A. Bendersky, "Cellular Microsegregation in Rapidly Solidified Ag-15 wt.%Cu Alloys," Rapidly Quenched Metals, Eds. S. Steeb and H. Warlimont (Elsevier Science Publishers B. V. 1985), 887-890.

8. M. C. Flemings, Solidification Processing (McGraw-Hill, 1974), 76-77.

9. W. J. Boettinger, "Growth Kinetic Limitations During Rapid Solidification," Rapidly Solidified Amorphous and Crystalline Alloys, Eds. B. H. Kear, B. C. Giessen and M. Cohen (Materials Research Society, Symposia Proceedings, Vol. 8, 1982), 15-31.

# SUBJECT INDEX

# AUTHOR INDEX